[illegible]out the A[illegible]

Titu Andreescu is the director of the American Mathematics Competitions, served as head coach of the USA International Mathematical Olympiad (IMO) Team, is Chair of the USA Mathematical Olympiad Committee, and is director of the Mathematical Olympiad Summer Program. [illegible] and Romania. [illegible] received the [illegible] Teacher Award from [illegible] Ministry of Education in 1983. [illegible] moving to the U.S., [illegible] was awarded the Edyth May Sliffe Award for Distinguished High School Mathematics Teaching from the MAA in 1994. [illegible] received a Certificate of Appreciation presented by the President of the MAA for "his outstanding service as coach of the USA Mathematical Olympiad Program in preparing the US team for its perfect performance in Hong Kong at the 1994 IMO."

Zuming Feng graduated with a Ph.D. [illegible] from Johns Hopkins University with emphasis on Algebraic Number Theory and Elliptic Curves. He teaches at Phillips Exeter Academy. [illegible] serves as a coach of the USA International Mathematical Olympiad (IMO) Team, a member of the USA Mathematical Olympiad Committee, an assistant director of the USA Mathematical Olympiad Summer Program. He received the Edyth May Sliffe Award for Distinguished High School Mathematics Teaching from the MAA in 1996 and in 2002.

About the Authors

Titu Andreescu is the Director of the American Mathematics Competitions, serves as Head Coach of the USA International Mathematical Olympiad (IMO) Team, is Chair of the USA Mathematical Olympiad Committee, and is Director of the Mathematical Olympiad Summer Program. Originally from Romania, Prof. Andreescu received the Distinguished Teacher Award from the Romanian Ministry of Education in 1983; then, after moving to the USA, he was awarded the Edyth May Sliffe Award for Distinguished High School Mathematics Teaching from the MAA in 1994. In addition, he received a Certificate of Appreciation presented by the President of the MAA for "his outstanding service as coach of the USA Mathematical Olympiad Program in preparing the USA team for its perfect performance in Hong Kong at the 1994 IMO."

Zuming Feng graduated with a Ph.D degree from Johns Hopkins University with emphasis on Algebraic Number Theory and Elliptic Curves. He teaches at Phillips Exeter Academy. He also serves as a coach of the USA International Mathematical Olympiad (IMO) Team, a member of the USA Mathematical Olympiad Committee, and an assistant director of the USA Mathematical Olympiad Summer Program. He received the Edyth May Sliffe Awards for Distinguished High School Mathematics Teaching from the MAA in 1996 and in 2002.

Titu Andreescu
Zuming Feng

102 Combinatorial Problems

From the Training of the USA IMO Team

Birkhäuser
Boston • Basel • Berlin

Titu Andreescu
American Mathematics Competitions
University of Nebraska
Lincoln, NE 68588
U.S.A.

Zuming Feng
Phillips Exeter Academy
Department of Mathematics
Exeter, NH 03833
U.S.A.

Library of Congress Cataloging-in-Publication Data

Andreescu, Titu, 1956-
102 combinatorial problems : from the training of the USA IMO team / Titu Andreescu and Zuming Feng.
p. cm.
Includes bibliographical references.
ISBN 0-8176-4317-6 (alk. paper) – ISBN 3-7643-4317-6 (alk. paper)
1. Combinatorial analysis–Problems, exercises, etc. 2. U.S.A. Mathematical Olympiad. I. Title: One hundred and two combinatorial problems. II. Feng, Zuming. III. Title.

QA164.A55 2002
511'.6–dc21

2002028382
CIP

AMS Subject Classifications: 05-XX

Printed on acid-free paper.

Birkhäuser 

ISBN 0-8176-4317-6 SPIN 10892938
ISBN 3-7643-4317-6

Reformatted from the authors' files by TEXniques, Inc., Cambridge, MA.
Printed in the United States of America.

9 8 7 6 5 4 3 2 1

Birkhäuser Boston • Basel • Berlin
A member of BertelsmannSpringer Science+Business Media GmbH

Contents

Preface and Introduction

This book contains 102 highly selected problems used in the training and testing of the USA International Mathematical Olympiad (IMO) team. It is not a collection of very difficult, impenetrable questions. Instead, the book gradually builds students' combinatorial skills and techniques. It aims to broaden a student's view of mathematics in preparation for possible participation in mathematical competitions. Problem-solving tactics and strategies further stimulate interest and confidence in combinatorics and other areas of mathematics.

Introduction

In the United States of America, the selection process leading to participation in the International Mathematical Olympiad (IMO) consists of a series of national contests called the American Mathematics Contest 10 (AMC 10), the American Mathematics Contest 12 (AMC 12), the American Invitational Mathematics Examination (AIME), and the United States of America Mathematical Olympiad (USAMO). Participation in the AIME and the USAMO is by invitation only, based on performance in the preceding exams of the sequence. The Mathematical Olympiad Summer Program (MOSP) is a four-week intensive training program for approximately one hundred very promising students who have risen to the top in American Mathematics Competitions. The six students representing the United States of America in the IMO are selected on the basis of their USAMO scores and further testing that takes place during the MOSP. Throughout the MOSP, full days of classes and extensive problem sets give students thorough preparation in several important areas of mathematics. These topics include combinatorial arguments and identities, generating functions, graph theory, recursive relations, sums and products, probability, number theory, polynomials, theory of equations, complex numbers in geometry, algorithmic proofs, combinatorial and advanced geometry, functional equations, and classical inequalities.

Olympiad-style exams consist of several challenging essay problems. Correct solutions often require deep analysis and careful argument. Olympiad questions can seem impenetrable to the novice, yet most can be solved with elementary high school mathematics techniques, cleverly applied.

Here is some advice for students who attempt the problems that follow.

- Take your time! Very few contestants can solve all the given problems.
- Try to make connections between problems. An important theme of this work is: all important techniques and ideas featured in the book appear more than once!
- Olympiad problems don't "crack" immediately. Be patient. Try different approaches. Experiment with simple cases. In some cases, working backward from the desired result is helpful.
- Even if you can solve a problem, read the solutions. They may contain some ideas that did not occur in your solutions, and they may discuss strategic and tactical approaches that can be used elsewhere. The solutions are also models of elegant presentation that you should emulate, but they often obscure the torturous process of investigation, false starts, inspiration, and attention to detail that led to them. When you read the solutions, try to reconstruct the thinking that went into them. Ask yourself: "What were the key ideas?" "How can I apply these ideas further?"
- Go back to the original problem later, and see if you can solve it in a different way. Many of the problems have multiple solutions, but not all are outlined here.
- Meaningful problem solving takes practice. Don't get discouraged if you have trouble at first. For additional practice, use the books on the reading list.

Acknowledgments

Thanks to Po-Ling Loh, Po-Ru Loh, and Tim Perrin who helped with typesetting, proofreading and preparing solutions.

Many problems are either inspired by or adapted from mathematical contests in different countries and from the following journals:

High-School Mathematics, China
Revista Matematică Timişoara, Romania

We did our best to cite all the original sources of the problems in the solution section. We express our deepest appreciation to the original proposers of the problems.

Abbreviations and Notations

Abbreviations

AHSME	American High School Mathematics Examination
AIME	American Invitational Mathematics Examination
AMC10	American Mathematics Contest 10
AMC12	American Mathematics Contest 12, which replaces AHSME
ARML	American Regional Mathematics League
IMO	International Mathematical Olympiad
USAMO	United States of America Mathematical Olympiad
MOSP	Mathematical Olympiad Summer Program
Putnam	The William Lowell Putnam Mathematical Competition
St. Petersburg	St. Petersburg (Leningrad) Mathematical Olympiad

Notations for Numerical Sets and Fields

$\mathbb{Z}$	the set of integers
$\mathbb{Z}_n$	the set of integers modulo n
$\mathbb{N}$	the set of positive integers
$\mathbb{N}_0$	the set of nonnegative integers
$\mathbb{Q}$	the set of rational numbers
$\mathbb{Q}^+$	the set of positive rational numbers
$\mathbb{Q}^0$	the set of nonnegative rational numbers
$\mathbb{Q}^n$	the set of n-tuples of rational numbers
$\mathbb{R}$	the set of real numbers
$\mathbb{R}^+$	the set of positive real numbers
$\mathbb{R}^0$	the set of nonnegative real numbers
$\mathbb{R}^n$	the set of n-tuples of real numbers
$\mathbb{C}$	the set of complex numbers
$[x^n](p(x))$	the coefficient of the term x^n in the polynomial $p(x)$

Notations for Sets, Logic, and Geometry

$\|A\|$	the number of elements in set A
$A \subset B$	A is a proper subset of B
$A \subseteq B$	A is a subset of B
$A \setminus B$	A without B
$A \cap B$	the intersection of sets A and B
$A \cup B$	the union of sets A and B
$a \in A$	the element a belongs to the set A

1

Introductory Problems

1. Mr. and Mrs. Zeta want to name their baby Zeta so that its monogram (first, middle, and last initials) will be in alphabetical order with no letters repeated. How many such monograms are possible?

2. The student lockers at Olympic High are numbered consecutively beginning with locker number 1. The plastic digits used to number the lockers cost two cents a piece. Thus, it costs two cents to label locker number 9 and four cents to label locker number 10. If it costs $137.94 to label all the lockers, how many lockers are there at the school?

3. Let n be an odd integer greater than 1. Prove that the sequence

$$\binom{n}{1}, \binom{n}{2}, \dots, \binom{n}{\frac{n-1}{2}}$$

contains an odd number of odd numbers.

4. How many positive integers not exceeding 2001 are multiples of 3 or 4 but not 5?

5. Let

$$x = .123456789101112\dots998999,$$

where the digits are obtained by writing the integers 1 through 999 in order. Find the 1983rd digit to the right of the decimal point.

6. Twenty five boys and twenty five girls sit around a table. Prove that it is always possible to find a person both of whose neighbors are girls.

7. At the end of a professional bowling tournament, the top 5 bowlers have a play-off. First #5 bowls #4. The loser receives 5th prize and the winner bowls #3 in another game. The loser of this game receives 4th prize and the winner bowls #2. The loser of this game receives 3rd prize and the winner bowls #1. The winner of this game gets 1st prize and the loser gets 2nd prize. In how many orders can bowlers #1 through #5 receive the prizes?

8. A spider has one sock and one shoe for each of its eight legs. In how many different orders can the spider put on its socks and shoes, assuming that, on each leg, the sock must be put on before the shoe?

9. A drawer in a darkened room contains 100 red socks, 80 green socks, 60 blue socks and 40 black socks. A youngster selects socks one at a time from the drawer but is unable to see the color of the socks drawn. What is the smallest number of socks that must be selected to guarantee that the selection contains at least 10 pairs? (A pair of socks is two socks of the same color. No sock may be counted in more than one pair.)

10. Given a rational number, write it as a fraction in lowest terms and calculate the product of the resulting numerator and denominator. For how many rational numbers between 0 and 1 will 20! be the resulting product?

11. Determine the number of ways to choose five numbers from the first eighteen positive integers such that any two chosen numbers differ by at least 2.

12. In a room containing N people, $N > 3$, at least one person has not shaken hands with everyone else in the room. What is the maximum number of people in the room that could have shaken hands with everyone else?

13. Find the number of ordered quadruples (x_1, x_2, x_3, x_4) of positive odd integers that satisfy $x_1 + x_2 + x_3 + x_4 = 98$.

14. Finitely many cards are placed in two stacks, with more cards in the left stack than in the right. Each card has one or more distinct names written on it, although different cards may share some names. For each name, we define a *shuffle* by moving every card that has that name written on it to the opposite stack. Prove that it is always possible to end up with more cards in the right stack by picking several distinct names, and doing in turn the shuffle corresponding to each name.

15. For how many pairs of consecutive integers in the set

$$\{1000, 1001, 1002, \ldots, 2000\}$$

is no carrying required when the two integers are added?

16. Nine chairs in a row are to be occupied by six students and Professors Alpha, Beta, and Gamma. These three professors arrive before the six students and decide to choose their chairs so that each professor will be between two students. In how many ways can Professors Alpha, Beta, and Gamma choose their chairs?

17. Prove that among any 16 distinct positive integers not exceeding 100 there are four different ones, a, b, c, d, such that $a + b = c + d$.

18. A child has a set of 96 distinct blocks. Each block is one of 2 materials *(plastic, wood)*, 3 sizes *(small, medium, large)*, 4 colors *(blue, green, red, yellow)*, and 4 shapes *(circle, hexagon, square, triangle)*. How many blocks in the set are different from the *"plastic medium red circle"* in exactly two ways? (The *"wood medium red square"* is such a block.)

19. Call a 7-digit telephone number $d_1d_2d_3 - d_4d_5d_6d_7$ *memorable* if the prefix sequence $d_1d_2d_3$ is exactly the same as either of the sequences $d_4d_5d_6$ or $d_5d_6d_7$ (possibly both). Assuming that each d_i can be any of the ten decimal digits $0, 1, 2, \dots, 9$, find the number of different memorable telephone numbers.

20. Two of the squares of a 7×7 checkerboard are painted yellow, and the rest are painted green. Two color schemes are equivalent if one can be obtained from the other by applying a rotation in the plane of the board. How many inequivalent color schemes are possible?

21. In how many ways can one arrange the numbers 21, 31, 41, 51, 61, 71, and 81 such that the sum of every four consecutive numbers is divisible by 3?

22. Let S be a set with six elements. In how many different ways can one select two not necessarily distinct subsets of S so that the union of the two subsets is S? The order of the selection does not matter; for example the pair of subsets $\{a, c\}, \{b, c, d, e, f\}$ represents the same selection as the pair $\{b, c, d, e, f\}, \{a, c\}$.

23. A set of positive numbers has the *triangle property* if it has three distinct elements that are the lengths of the sides of a triangle whose area is positive. Consider sets $\{4, 5, 6, \dots, n\}$ of consecutive positive integers, all of whose ten-element subsets have the triangle property. What is the largest possible value of n?

24. Let A and B be disjoint sets whose union is the set of natural numbers. Show that for every natural number n there exist distinct $a, b > n$ such that

$$\{a, b, a + b\} \subseteq A \quad \text{or} \quad \{a, b, a + b\} \subseteq B.$$

25. The increasing sequence $1, 3, 4, 9, 10, 12, 13, \ldots$ consists of all those positive integers which are powers of 3 or sums of distinct powers of 3. Find the 100þ term of this sequence (where 1 is the 1st term, 3 is the 2nd term, and so on).

26. Every card in a deck has a picture of one shape — circle, square, or triangle, which is painted in one of three colors — red, blue, or green. Furthermore, each color is applied in one of three shades — light, medium, or dark. The deck has 27 cards, with every shape-color-shade combination represented. A set of three cards from the deck is called *complementary* if all of the following statements are true:

    (a) Either each of the three cards has a different shape or all three of the cards have the same shape.

    (b) Either each of the three cards has a different color or all three of the cards have the same color.

    (c) Either each of the three cards has a different shade or all three of the cards have the same shade.

    How many different complementary three-card sets are there?

27. At a math camp, all m students share exactly one common friend, $m \geq 3$. (If A is a friend of B, then B is a friend of A. Also, A is not his own friend.) Suppose person P has the largest number of friends. Determine what that number is.

28. Suppose that 7 boys and 13 girls line up in a row. Let S be the number of places in the row where a boy and a girl are standing next to each other. For example, for the row $GBBGGGBGBGGGBGBGGBGG$ we have $S = 12$. Find the average value of S (if all possible orders of these 20 people are considered).

29. A bored student walks down a hall that contains a row of closed lockers, numbered 1 to 1024. He opens the locker numbered 1, and then alternates between skipping and opening each closed locker thereafter. When he reaches the end of the hall, the student turns around and starts back. He opens the first closed locker he encounters, and then alternates between skipping and opening each closed locker thereafter. The student continues wandering back and forth in this manner until every locker is open. What is the number of the last locker he opens?

30. Let $n = 2^{31}3^{19}$. How many positive integer divisors of n^2 are less than n but do not divide n?

31. In an arena, each row has 199 seats. One day, 1990 students are coming to attend a soccer match. It is only known that at most 39 students are from the same school. If students from the same school must sit in the same row, determine the minimum number of rows that must be reserved for these students.

32. Let $T = \{9^k \mid k \text{ is an integer}, 0 \leq k \leq 4000\}$. Given that 9^{4000} has 3817 digits and that its first (leftmost) digit is 9, how many elements of T have 9 as their leftmost digit?

33. For what values of $n \geq 1$ do there exist a number m that can be written in the form $a_1 + \cdots + a_n$ (with $a_1 \in \{1\}, a_2 \in \{1, 2\}, \ldots, a_n \in \{1, \ldots, n\}$) in $(n-1)!$ or more ways?

34. Let the sum of a set of numbers be the sum of its elements. Let S be a set of positive integers, none greater than 15. Suppose no two disjoint subsets of S have the same sum. What is the largest sum a set S with these properties can have?

35. There are at least four candy bars in n ($n \geq 4$) boxes. Each time, Mr. Fat is allowed to pick two boxes, take one candy bar from each of the two boxes, and put those candy bars into a third box. Determine if it is always possible to put all the candy bars into one box.

36. Determine, with proof, if it is possible to arrange $1, 2, \ldots, 1000$ in a row such that the average of any pair of distinct numbers is not located in between the two numbers.

37. Let $A_1A_2 \ldots A_{12}$ be a regular dodecagon with O as its center. Triangular regions OA_iA_{i+1}, $1 \leq i \leq 12$ (and $A_{13} = A_1$) are to be colored red, blue, green, or yellow such that adjacent regions are colored in different colors. In how many ways can this be done?

38. There are $2n$ people at a party. Each person has an even number of friends at the party. (Here friendship is a mutual relationship.) Prove that there are two people who have an even number of common friends at the party.

39. How many different 4×4 arrays whose entries are all $1's$ and $-1's$ have the property that the sum of the entries in each row is 0 and the sum of the entries in each column is 0?

40. A square of dimensions $(n-1) \times (n-1)$ is divided into $(n-1)^2$ unit squares in the usual manner. Each of the n^2 vertices of these squares is to be colored red or blue. Find the number of different colorings such that each unit square has exactly two red vertices. (Two coloring schemes are regarded as different if at least one vertex is colored differently in the two schemes.)

41. Sixty-four balls are separated into several piles. At each step we are allowed to apply the following operation. Pick two piles, say pile $\mathcal{A}$ with p balls and pile $\mathcal{B}$ with q balls and $p \geq q$, and then remove q balls from pile $\mathcal{A}$ and put them in pile $\mathcal{B}$. Prove that it is possible to put all the balls into one pile.

42. A game of solitaire is played with a finite number of nonnegative integers. On the first move the player designates one integer as *large*, and replaces another integer by any nonnegative integer strictly smaller than the designated large integer. On subsequent steps the play is similar, except that the integer replaced must be the one designated as large on the previous play. Prove that in some finite number of steps the play must end.

43. Given $S \subseteq \{1, 2, \ldots, n\}$, we are allowed to modify it in any one of the following ways:

    (a) if $1 \notin S$, add the element 1;

    (b) if $n \in S$, delete the element n;

    (c) for $1 \leq r \leq n-1$, if $r \in S$ and $r+1 \notin S$, delete the element r and add the element $r+1$.

    Suppose that it is possible by such modifications to obtain a sequence

    $$\emptyset \to \{1\} \to \{2\} \to \cdots \to \{n\},$$

    starting with $\emptyset$ and ending with $\{n\}$, in which each of the 2^n subsets of $\{1, 2, \ldots, n\}$ appears exactly once. Prove that $n = 2^m - 1$ for some m.

44. There are 2001 coins on a table. For $i = 1, 2, \ldots, 2001$ in succession, one must turn over exactly i coins. Prove that it is always possible either to make all of the coins face up or to make all of the coins face down, but not both.

45. For $\{1, 2, \ldots, n\}$ and each of its nonempty subsets a unique *alternating sum* is defined as follows: Arrange the numbers in the subset in decreasing order and then, beginning with the largest, alternately add and subtract successive numbers. (For example, the alternating sum for $\{1, 2, 4, 6, 9\}$ is $9 - 6 + 4 - 2 + 1 = 6$ and for $\{5\}$ it is simply 5.) Find the sum of all such alternating sums for $n = 7$.

46. In a game of *Chomp*, two players alternately take "bites" from a 5-by-7 grid of unit squares. To take a bite, the player chooses one of the remaining squares, then removes ("eats") all squares found in the quadrant defined by the left edge (extended upward) and the lower edge (extended rightward) of the chosen square. For example, the bite determined by the shaded square in

the diagram would remove the shaded square and the four squares marked by $\times$.

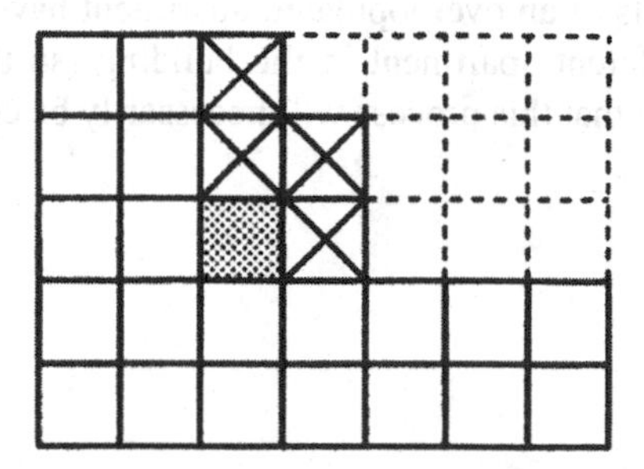

(The squares with two or more dotted edges have been removed from the original board in previous moves.) The object of the game is to make one's opponent take the last bite. The diagram shows one of the many subsets of the set of 35 unit squares that can occur during the game of Chomp. How many different subsets are there in all? Include the full board and the empty board in your count.

47. Each square of a 1998×2002 chess board contains either 0 or 1 such that the total number of squares containing 1 is odd in each row and each column. Prove that the number of white unit squares containing 1 is even.

48. Let S be a subset of $\{1, 2, 3, \ldots, 1989\}$ such that no two members of S differ by 4 or 7. What is the largest number of elements S can have?

49. A class of fifteen boys and fifteen girls is seated around a round table. Their teacher wishes to pair up the students and hand out fifteen tests—one test to each pair.

    As the teacher is preparing to select the pairs and hand out the tests, he wonders to himself: "How many seating arrangements would allow me to match up boy/girl pairs sitting next to each other without having to ask any student to change his or her seat?" Answer the teacher's question. (Two seating arrangements are regarded as being the same if one can be obtained from the other by a rotation.)

50. Two squares on an 8×8 chessboard are called *touching* if they have at least one common vertex. Determine if it is possible for a king to begin in some square and visit all the squares exactly once in such a way that all moves except the first are made into squares touching an even number of squares already visited.

51. A total of 119 residents live in a building with 120 apartments. We call an apartment *overpopulated* if there are at least 15 people living there. Every day the inhabitants of an overpopulated apartment have a quarrel and each goes off to a different apartment in the building (so they can avoid each other☺). Is it true that this process will necessarily be completed someday?

2

Advanced Problems

1. In a tournament each player played exactly one game against each of the other players. In each game the winner was awarded 1 point, the loser got 0 points, and each of the two players earned 1/2 point if the game was a tie. After the completion of the tournament, it was found that exactly half of the points earned by each player were earned in games against the ten players with the least number of points. (In particular, each of the ten lowest scoring players earned half of her/his points against the other nine of the ten). What was the total number of players in the tournament?

2. Let n be an odd integer greater than 1. Find the number of permutations p of the set $\{1, 2, \ldots, n\}$ for which

$$|p(1) - 1| + |p(2) - 2| + \cdots + |p(n) - n| = \frac{n^2 - 1}{2}.$$

3. In a sequence of coin tosses one can keep a record of the number of instances when a tail is immediately followed by a head, a head is immediately followed by a head, etc. We denote these by TH, HH, etc. For example, in the sequence $HHTTHHHHTHHTTTT$ of 15 coin tosses we observe that there are five HH, three HT, two TH , and four TT subsequences. How many different sequences of 15 coin tosses will contain exactly two HH, three HT, four TH and five TT subsequences?

4. Let $A = (a_1, a_2, \ldots, a_{2001})$ be a sequence of positive integers. Let m be the number of 3-element subsequences (a_i, a_j, a_k) with $1 \le i < j < k \le 2001$, such that $a_j = a_i + 1$ and $a_k = a_j + 1$. Considering all such sequences A, find the greatest value of m.

5. Twenty-three people of positive integral weights decide to play football. They select one person as referee and then split up into two 11-person teams of equal total weights. It turns out that no matter who the referee this can always be done. Prove that all 23 people have equal weights.

6. Determine the smallest integer n, $n \ge 4$, for which one can choose four different numbers a, b, c, d from any n distinct integers such that $a+b-c-d$ is divisible by 20.

7. A mail carrier delivers mail to the nineteen houses on the east side of Elm Street. The carrier notes that no two adjacent houses ever get mail on the same day, but that there are never more than two houses in a row that get no mail on the same day. How many different patterns of mail delivery are possible?

8. For $i = 1, 2, \ldots, 11$, let M_i be a set of five elements, and assume that for every $1 \le i < j \le 11$, $M_i \cap M_j \ne \emptyset$. Let m be the largest number for which there exist $M_{i_1}, \ldots, M_{i_m}$ among the chosen sets with $\cap_{k=1}^{m} M_{i_k} \ne \emptyset$. Find the minimum value of m over all possible initial choices of M_i.

9. Define a *domino* to be an ordered pair of *distinct* positive integers. A *proper sequence* of dominos is a list of distinct dominos in which the first coordinate of each pair after the first equals the second coordinate of the immediately preceding pair, and in which (i, j) and (j, i) do not *both* appear for any i and j. Let D_{40} be the set of all dominos whose coordinates are no larger than 40. Find the length of the longest proper sequence of dominos that can be formed using the dominos of D_{40}.

10. Find the number of subsets of $\{1, \ldots, 2000\}$, the sum of whose elements is divisible by 5.

11. Let X be a finite set of positive integers and A a subset of X. Prove that there exists a subset B of X such that A equals the set of elements of X which divide an odd number of elements of B.

12. A stack of 2000 cards is labeled with the integers from 1 to 2000, with different integers on different cards. The cards in the stack are not in numerical order. The top card is removed from the stack and placed on the table, and the next card in the stack is moved to the bottom of the stack. The new top card is removed from the stack and placed on the table, to the right of the card

already there, and the next card in the stack is moved to the bottom of the stack. This process—placing the top card to the right of the cards already on the table and moving the next card in the stack to the bottom of the stack—is repeated until all cards are on the table. It is found that, reading left to right, the labels on the cards are now in ascending order: $1, 2, 3, \ldots, 1999, 2000$. In the original stack of cards, how many cards were above the card labeled 1999?

13. Form a 2000×2002 screen with unit screens. Initially, there are more than 1999×2001 unit screens which are *on*. In any 2×2 screen, as soon as there are 3 unit screens which are *off*, the 4þ screen turns off automatically. Prove that the whole screen can never be totally off.

14. In an office, at various times during the day, the boss gives the secretary a letter to type, each time putting the letter on top of the pile in the secretary's inbox. When there is time, the secretary takes the top letter off the pile and types it. There are nine letters to be typed during the day, and the boss delivers them in the order $1, 2, 3, 4, 5, 6, 7, 8, 9$. While leaving for lunch, the secretary tells a colleague that letter 8 has already been typed, but says nothing else about the morning's typing. The colleague wonders which of the nine letters remain to be typed after lunch and in what order they will be typed. Based upon the above information, how many such *after lunch typing orders* are possible? (That there are no letters left to be typed is one of the possibilities.)

15. Let n be a positive integer. Prove that

$$\sum_{k=0}^{n} 2^k \binom{n}{k} \binom{n-k}{\lfloor (n-k)/2 \rfloor} = \binom{2n+1}{n}.$$

16. Let m and n be positive integers. Suppose that a given rectangle can be tiled by a combination of horizontal $1 \times m$ strips and vertical $n \times 1$ strips. Prove that it can be tiled using only one of the two types.

17. Given an initial sequence $a_1, a_2, \ldots, a_n$ of real numbers, we perform a series of steps. At each step, we replace the current sequence $x_1, x_2, \ldots, x_n$ with $|x_1 - a|, |x_2 - a|, \ldots, |x_n - a|$ for some a. For each step, the value of a can be different.

    (a) Prove that it is always possible to obtain the null sequence consisting of all 0's.

    (b) Determine with proof the minimum number of steps required, regardless of initial sequence, to obtain the null sequence.

18. The sequence $\{a_n\}_{n\geq 1}$ satisfies the conditions $a_1 = 0, a_2 = 1$,

$$a_n = \frac{1}{2}na_{n-1} + \frac{1}{2}n(n-1)a_{n-2} + (-1)^n\left(1 - \frac{n}{2}\right),$$

$n \geq 3$. Determine the explicit form of

$$\begin{aligned} f_n &= a_n + 2\binom{n}{1}a_{n-1} + 3\binom{n}{2}a_{n-2} \\ &\quad + \cdots + (n-1)\binom{n}{n-2}a_2 + n\binom{n}{n-1}a_1. \end{aligned}$$

19. For a set A, let $|A|$ and $s(A)$ denote the number of the elements in A and the sum of elements in A, respectively. (If $A = \emptyset$, then $|A| = s(A) = 0$.) Let S be a set of positive integers such that

    (a) there are two numbers $x, y \in S$ with $\gcd(x, y) = 1$;

    (b) for any two numbers $x, y \in S$, $x + y \in S$.

    Let T be the set of all positive integers not in S. Prove that $s(T) \leq |T|^2 < \infty$.

20. In a forest each of 9 animals lives in its own cave, and there is exactly one separate path between any two of these caves. Before the election for Forest Gump, King of the Forest, some of the animals make an election campaign. Each campaign-making animal—$\mathcal{FGC}$ (Forest Gump candidate)—visits each of the other caves exactly once, uses only the paths for moving from cave to cave, never turns from one path to another between the caves, and returns to its own cave at the end of the campaign. It is also known that no path between two caves is used by more than one $\mathcal{FGC}$. Find the maximum possible number of $\mathcal{FGC}$'s.

21. For a sequence $A_1, \ldots, A_n$ of subsets of $\{1, \ldots, n\}$ and a permutation π of $S = \{1, \ldots, n\}$, we define the diagonal set

$$D_\pi(A_1, A_2, \ldots, A_n) = \{i \in S \mid i \notin A_{\pi(i)}\}.$$

    What is the maximum possible number of distinct sets which can occur as diagonal sets for a single choice of $A_1, \ldots, A_n$?

22. A subset M of $\{1, 2, 3, \ldots, 15\}$ does not contain three elements whose product is a perfect square. Determine the maximum number of elements in M.

23. Find all finite sequences $(x_0, x_1, \ldots, x_n)$ such that for every j, $0 \leq j \leq n$, x_j equals the number of times j appears in the sequence.

24. Determine if it is possible to partition the set of positive integers into sets $\mathcal{A}$ and $\mathcal{B}$ such that $\mathcal{A}$ does not contain any 3-element arithmetic sequence and $\mathcal{B}$ does not contain any infinite arithmetic sequence.

25. Consider the set T_5 of 5-digit positive integers whose decimal representations are permutations of the digits 1, 2, 3, 4, 5. Determine if it is possible to partition T_5 into sets A and B such that the sum of the squares of the elements in A is equal to the corresponding sum for B.

26. Let n be a positive integer. Find the number of polynomials $P(x)$ with coefficients in $\{0, 1, 2, 3\}$ such that $P(2) = n$.

27. Let n and k be positive integers such that $\frac{1}{2}n < k \leq \frac{2}{3}n$. Find the least number m for which it is possible to place each of m pawns on a square of an $n \times n$ chessboard so that no column or row contains a block of k adjacent unoccupied squares.

28. In a soccer tournament, each team plays another team exactly once and receives 3 points for a win, 1 point for a draw, and 0 points for a loss. After the tournament, it is observed that there is a team which has earned both the most total points and won the *fewest* games. Find the smallest number of teams in the tournament for which this is possible.

29. Let $a_1, \ldots, a_n$ be the first row of a triangular array with $a_i \in \{0, 1\}$. Fill in the second row $b_1, \ldots, b_{n-1}$ according to the rule $b_k = 1$ if $a_k \neq a_{k+1}$, $b_k = 0$ if $a_k = a_{k+1}$. Fill in the remaining rows similarly. Determine with proof the maximum possible number of 1's in the resulting array.

30. There are 10 cities in the *Fatland*. Two airlines control all of the flights between the cities. Each pair of cities is connected by exactly one flight (in both directions). Prove that one airline can provide two traveling cycles with each cycle passing through an odd number of cities and with no common cities shared by the two cycles.

31. Suppose that each positive integer not greater than $n(n^2 - 2n + 3)/2$, $n \geq 2$, is colored one of two colors (red or blue). Show that there must be a monochromatic n-term sequence $a_1 < a_2 < \cdots < a_n$ satisfying

$$a_2 - a_1 \leq a_3 - a_2 \leq \cdots \leq a_n - a_{n-1}.$$

32. The set $\{1, 2, \ldots, 3n\}$ is partitioned into three sets A, B, and C with each set containing n numbers. Determine with proof if it is always possible to choose one number out of each set so that one of these numbers is the sum of the other two.

33. Assume that each of the 30 MOPpers has exactly one favorite chess variant and exactly one favorite classical inequality. Each MOPper lists this information on a survey. Among the survey responses, there are exactly 20 different favorite chess variants and exactly 10 different favorite inequalities. Let n be the number of MOPpers M such that the number of MOPpers who listed M's favorite inequality is greater than the number of MOPpers who listed M's favorite chess variant. Prove that $n \geq 11$.

34. Starting from a triple (a, b, c) of nonnegative integers, a *move* consists of choosing two of them, say x and y, and replacing one of them by either $x+y$ or $|x-y|$. For example, one can go from $(3, 5, 7)$ to $(3, 5, 4)$ in one move. Prove that there exists a constant $r > 0$ such that whenever a, b, c, n are positive integers with $a, b, c < 2^n$, there is a sequence of at most rn moves transforming (a, b, c) into (a', b', c') with $a'b'c' = 0$.

35. A rectangular array of numbers is given. In each row and each column, the sum of all the numbers is an integer. Prove that each nonintegral number x in the array can be changed into either $\lceil x \rceil$ or $\lfloor x \rfloor$ so that the row-sums and the column-sums remain unchanged. (Note that $\lceil x \rceil$ is the least integer greater than or equal to x, while $\lfloor x \rfloor$ is the greatest integer less than or equal to x.)

36. A finite set of (distinct) positive integers is called a *DS-set* if each of the integers divides the sum of them all. Prove that every finite set of positive integers is a subset of some DS-set.

37. Twelve musicians $M_1, M_2, \ldots, M_{12}$ gather at a week-long chamber music festival. Each day, there is one scheduled concert and some of the musicians play while the others listen as members of the audience. For $i = 1, 2, \ldots, 12$, let t_i be the number of concerts in which musician M_i plays, and let $t = t_1 + t_2 + \cdots + t_{12}$. Determine the minimum value of t such that it is possible for each musician to listen, as a member of the audience, to all the other musicians.

38. An $m \times n$ array is filled with the numbers $\{1, 2, \ldots n\}$, each used exactly m times. Show that one can always permute the numbers within columns to arrange that each row contains every number $\{1, 2, \ldots, n\}$ exactly once.

39. Let the set $U = \{1, 2, \ldots, n\}$, where $n \geq 3$. A subset S of U is said to be *split* by an arrangement of the elements of U if an element not in S occurs in the arrangement somewhere between two elements of S. For example, 13542 splits $\{1, 2, 3\}$ but not $\{3, 4, 5\}$. Prove that for any $n-2$ subsets of U, each containing at least 2 and at most $n-1$ elements, there is an arrangement of the elements of U which splits all of them.

40. A pile of n pebbles is placed in a vertical column. This configuration is modified according to the following rules. A pebble can be moved if it is at the top of a column which contains at least two more pebbles than the column immediately to its right. (If there are no pebbles to the right, think of this as a column with 0 pebbles.) At each stage, choose a pebble from among those that can be moved (if there are any) and place it at the top of the column to its right. If no pebbles can be moved, the configuration is called a *final configuration*. For each n, show that, no matter what choices are made at each stage, the final configuration obtained is unique. Describe that configuration in terms of n.

41. Let B_n be the set of all binary strings of length n. Given two strings $(a_i)_{i=1}^n$ and $(b_i)_{i=1}^n$, define the distance between the strings as

$$d((a_i), (b_i)) = \sum_{k=1}^{n} |a_i - b_i|.$$

Let C_n be a subset of B_n. The set C_n is called a *perfect error correcting code (PECC) of length n and tolerance m* if for each string (b_i) in B_n there is a unique string (c_i) in C_n with $d((b_i), (c_i)) \leq m$. Prove that there is no PECC of length 90 and tolerance 2.

42. Determine if it is possible to arrange the numbers $1, 1, 2, 2, \ldots, n, n$ such that there are j numbers between two j's, $1 \leq j \leq n$, when $n = 2000$, $n = 2001$, and $n = 2002$. (For example, for $n = 4$, 41312432 is such an arrangement.)

43. Let k, m, n be integers such that $1 < n \leq m - 1 \leq k$. Determine the maximum size of a subset S of the set $\{1, 2, \ldots, k\}$ such that no n distinct elements of S add up to m.

44. A nondecreasing sequence $s_0, s_1, \ldots$ of nonnegative integers is said to be *superadditive* if $s_{i+j} \geq s_i + s_j$ for all nonnegative integers i, j. Suppose $\{s_n\}$ and $\{t_n\}$ are two superadditive sequences, and let $\{u_n\}$ be a nondecreasing sequence with the property that each integer appears in $\{u_n\}$ as many times as in $\{s_n\}$ and $\{t_n\}$ combined. Show that $\{u_n\}$ is also superadditive.

45. The numbers from 1 to n^2, $n \geq 2$, are randomly arranged in the cells of an $n \times n$ unit square grid. For any pair of numbers situated on the same row or on the same column, the ratio of the greater number to the smaller one is calculated. The *characteristic* of the arrangement is the smallest of these $n^2(n-1)$ fractions. Determine the largest possible value of the characteristic.

46. For a set S, let $|S|$ denote the number of elements in S. Let A be a set with $|A| = n$, and let $A_1, A_2, \dots, A_n$ be subsets of A with $|A_i| \geq 2$, $1 \leq i \leq n$. Suppose that for each 2-element subset A' of A, there is a unique i such that $A' \subseteq A_i$. Prove that $A_i \cap A_j \neq \emptyset$ for all $1 \leq i < j \leq n$.

47. Suppose that $r_1, \dots, r_n$ are real numbers. Prove that there exists a set $S \subseteq \{1, 2, \dots, n\}$ such that

$$1 \leq |S \cap \{i, i+1, i+2\}| \leq 2,$$

for $1 \leq i \leq n-2$, and

$$\left| \sum_{i \in S} r_i \right| \geq \frac{1}{6} \sum_{i=1}^{n} |r_i|.$$

48. Let n, k, m be positive integers with $n > 2k$. Let S be a nonempty set of k-element subsets of $\{1, \dots, n\}$ with the property that every $(k+1)$-element subset of $\{1, \dots, n\}$ contains exactly m elements of S. Prove that S must contain every k-element subset of $\{1, \dots, n\}$.

49. A set T is called *even* if it has an even number of elements. Let n be a positive even integer, and let $S_1, S_2, \dots, S_n$ be even subsets of the set $S = \{1, 2, \dots, n\}$. Prove that there exist i and j, $1 \leq i < j \leq n$, such that $S_i \cap S_j$ is even.

50. Let $A_1, A_2, \dots, B_1, B_2, \dots$ be sets such that $A_1 = \emptyset$, $B_1 = \{0\}$,

$$A_{n+1} = \{x + 1 \mid x \in B_n\}, \quad B_{n+1} = A_n \cup B_n - A_n \cap B_n,$$

for all positive integers n. Determine all the positive integers n such that $B_n = \{0\}$.

51. [Iran 1999] Suppose that $S = \{1, 2, \dots, n\}$ and that $A_1, A_2, \dots, A_k$ are subsets of S such that for every $1 \leq i_1, i_2, i_3, i_4 \leq k$, we have

$$|A_{i_1} \cup A_{i_2} \cup A_{i_3} \cup A_{i_4}| \leq n - 2.$$

Prove that $k \leq 2^{n-2}$.

3

Solutions to Introductory Problems

1. [AHSME 1989] Mr. and Mrs. Zeta want to name their baby Zeta so that its monogram (first, middle, and last initials) will be in alphabetical order with no letters repeated. How many such monograms are possible?

First Solution: The possible monograms are

$$ABZ, ACZ, \ldots, WXZ, WYZ, XYZ.$$

Any two-element subset of the first 25 letters of the alphabet, when used in alphabetical order, will produce a suitable monogram when combined with Z. For example $\{L, J\} = \{J, L\}$ will produce JLZ. Furthermore, to every suitable monogram there corresponds exactly one two-element subset of $\{A, B, C, \ldots, Y\}$. Thus, the answer is the number of two-element subsets that can be formed from a set of 25 letters, and there are $\binom{25}{2} = 300$ such subsets.

Second Solution: The last initial is fixed at Z. If the first initial is A, the second initial must be one of $B, C, D, \ldots, Y$, so there are 24 choices for the second. If the first initial is B, there are 23 choices for the second initial: $C, D, E, \ldots, Y$. Continuing in this way we see that the number of monograms is

$$24 + 23 + \ldots + 2 + 1.$$

Use the formula

$$1+2+\ldots+n=\frac{n(n+1)}{2}$$

to obtain the answer $\frac{24\cdot 25}{2}=300$.

2. [AHSME 1999] The student lockers at Olympic High are numbered consecutively beginning with locker number 1. The plastic digits used to number the lockers cost two cents a piece. Thus, it costs two cents to label locker number 9 and four cents to label locker number 10. If it costs \$137.94 to label all the lockers, how many lockers are there at the school?

Solution: The locker labeling requires $137.94/0.02=6897$ digits. Lockers 1 through 9 require 9 digits, lockers 10 through 99 require $2\cdot 90=180$ digits, and lockers 100 through 999 require $3\cdot 900=2700$ digits. Hence the remaining lockers require $6897-2700-180-9=4008$ digits, so there must be $4008/4=1002$ more lockers, each using four digits. In all, there are $1002+999=2001$ student lockers.

3. [Revista Matematică Timişoara] Let n be an odd integer greater than 1. Prove that the sequence

$$\binom{n}{1},\binom{n}{2},\ldots,\binom{n}{\frac{n-1}{2}}$$

contains an odd number of odd numbers.

Solution: The sum of the numbers in the given sequence equals

$$\frac{1}{2}\left[\binom{n}{1}+\binom{n}{2}+\cdots+\binom{n}{n-1}\right]=\frac{1}{2}(2^n-2)=2^{n-1}-1,$$

which is an odd number and the conclusion follows.

4. [AMC12 2001] How many positive integers not exceeding 2001 are multiples of 3 or 4 but not 5?

Solution: For integers not exceeding 2001, there are $\lfloor 2001/3\rfloor=667$ multiples of 3 and $\lfloor 2001/4\rfloor=500$ multiples of 4. The total, 1167, counts the $\lfloor 2001/12\rfloor=166$ multiples of 12 twice, so there are $1167-166=1001$ multiples of 3 or 4. From these we exclude the $\lfloor 2001/15\rfloor=133$ multiples of 15 and the $\lfloor 2001/20\rfloor=100$ multiples of 20, since these are multiples of 5. However, this excludes the $\lfloor 2001/60\rfloor=33$ multiples of 60 twice, so we must re-include them. The number of integers satisfying the conditions is $1001-133-100+33=801$.

5. [AHSME 1983] Let

$$x = .123456789101112\ldots998999,$$

where the digits are obtained by writing the integers 1 through 999 in order. Find the 1983rd digit to the right of the decimal point.

Solution: Look at the first 1983 digits, and let z denote the 1983rd digit. We may break this string of digits into three segments:

$$\underbrace{.123456789}_{A}\quad\underbrace{1011\ldots9899}_{B}\quad\underbrace{100101\ldots z}_{C}.$$

There are 9 digits in A, $2 \cdot 90 = 180$ in B, hence $1983 - 189 = 1794$ in C. Dividing 1794 by 3 we get 598 with remainder 0. Thus C consists of the first 598 3-digit integers. Since the first 3-digit integer is 100 (not 101 or 001), the 598th 3-digit integer is $598 + 99 = 697$. Thus $z = 7$.

6. Twenty five boys and twenty five girls sit around a table. Prove that it is always possible to find a person both of whose neighbors are girls.

First Solution: For the sake of contradiction we assume that there is a seating arrangement such that there is no one sitting in between two girls. We call a *block* any group of girls(boys) sitting next to each other and sandwiched by boys(girls) from both sides. By our assumption, each girl block has at most 2 girls and there are at least 2 boys in the gap between two consecutive girl blocks. Hence there are at least $\lceil 25/2 \rceil = 13$ girl blocks and at least 2×13 boys sitting in between the 13 gaps between girls blocks. But we only have 25 boys, a contradiction. Therefore our assumption was wrong and it is always possible to find someone sitting between two girls.

Second Solution: We again approach indirectly by assuming that there is a seating arrangement such that no one is sitting in between two girls. We further assume that they are sitting is positions $a_1, a_2, \ldots, a_{50}$ in a counterclockwise order (so a_{50} is next to a_1). Now we split them into two tables with seating orders $(a_1, a_3, a_5, \ldots, a_{49})$ and $(a_2, a_4, a_6, \ldots, a_{50})$, each in counterclockwise order. Then by our assumption, no girls are next to each other in the resulting two-seating arrangements. So there are at most 12 girls sitting around each new table for a total of at most 24 girls, a contradiction. Therefore our assumption was wrong and it is always possible to find someone sitting in between two girls.

7. [AHSME 1988] At the end of a professional bowling tournament, the top 5 bowlers have a play-off. First #5 bowls #4. The loser receives 5th prize and the winner bowls #3 in another game. The loser of this game receives receives 4th prize and the winner bowls #2. The loser of this game receives 3rd prize and the winner bowls #1. The winner of this game gets 1st prize and the loser gets 2nd prize. In how many orders can bowlers #1 through #5 receive the prizes?

   Solution: There are 4 games in every playoff, and each game has 2 possible outcomes. For each sequence of 4 outcomes, the prizes are awarded in a different way. Thus there are $2^4 = 16$ possible orders.

8. [AMC12 2001] A spider has one sock and one shoe for each of its eight legs. In how many different orders can the spider put on its socks and shoes, assuming that, on each leg, the sock must be put on before the shoe?

   Solution: Number the spider's legs from 1 through 8, and let a_k and b_k denote the sock and shoe that will go on leg k. A possible arrangement of the socks and shoes is a permutation of the sixteen symbols $a_1, b_1, \ldots, a_8, b_8$, in which a_k precedes b_k for $1 \leq k \leq 8$. There are 16! permutations of the sixteen symbols, and a_1 precedes b_1 in exactly half of these, or $16!/2$ permutations. Similarly, a_2 precedes b_2 in exactly half of those, or $16!/2^2$ permutations. Continuing, we can conclude that a_k precedes b_k for $1 \leq k \leq 8$ in exactly $16!/2^8$ permutations.

9. [AHSME 1986] A drawer in a darkened room contains 100 red socks, 80 green socks, 60 blue socks and 40 black socks. A youngster selects socks one at a time from the drawer but is unable to see the color of the socks drawn. What is the smallest number of socks that must be selected to guarantee that the selection contains at least 10 pairs? (A pair of socks is two socks of the same color. No sock may be counted in more than one pair.)

   First Solution: For any selection, at most one sock of each color will be left unpaired, and this happens if and only if an odd number of socks of that color is selected. Thus 24 socks suffice: at most 4 will be unpaired, leaving at least 20 in pairs. However, 23 will do! Since 23 is not the sum of four odd numbers, at most 3 socks out of 23 will be unpaired. On the other hand, 22 will *not* do: if the numbers of red, green, blue, and black socks are 5,5,5,7, then four are unpaired, leaving 9 pairs. Thus 23 is the minimum.

   Second Solution: Proceed inductively. If we require only one pair, then it suffices to select 5 socks. Moreover, selecting 4 socks doesn't guarantee a pair since we might select one sock of each color.

If we require two pairs, then it suffices to select 7 socks: any set of 7 socks must contain a pair; if we remove this pair, then the remaining 5 socks will contain a second pair as shown above. On the other hand, 6 socks might contain 3 greens, 1 black, 1 red and 1 blue – hence only one pair. Thus 7 socks is the smallest number to guarantee two pairs.

Similar reasoning shows that we must draw 9 socks to guarantee 3 pairs, and in general, $2p+3$ socks to guarantee p pairs. This formula is easily proved by mathematical induction. Thus 23 socks are needed to guarantee 10 pairs.

10. [AIME 1991] Given a rational number, write it as a fraction in lowest terms and calculate the product of the resulting numerator and denominator. For how many rational numbers between 0 and 1 will 20! be the resulting product?

Solution: For a fraction to be in lowest terms, its numerator and denominator must be relatively prime. Thus any prime factor that occurs in the numerator cannot occur in the denominator, and vice-versa. There are eight prime factors of 20!, namely 2, 3, 5, 7, 11, 13, 17, and 19. For each of these prime factors, one must decide only whether it occurs in the numerator or in the denominator. These eight decisions can be made in a total of $2^8 = 256$ ways. However, not all of the 256 resulting fractions will be less than 1. Indeed, they can be grouped into 128 pairs of reciprocals, each containing exactly one fraction less than 1. Thus the number of rational numbers with the desired property is 128.

11. Determine the number of ways to choose five numbers from the first eighteen positive integers such that any two chosen numbers differ by at least 2.

Solution: Let $a_1 < a_2 < a_3 < a_4 < a_5$ be the five chosen numbers. Consider the numbers $(b_1, b_2, b_3, b_4, b_5) = (a_1, a_2 - 1, a_3 - 2, a_4 - 3, a_5 - 4)$. Then b_1, b_2, b_3, b_4, b_5 are five distinct numbers from the first fourteen positive integers. Conversely, from any five distinct numbers $b_1 < b_2 < b_3 < b_4 < b_5$ we can reconstruct $(a_1, a_2, a_3, a_4, a_5) = (b_1, b_2 + 1, b_3 + 2, b_4 + 3, b_5 + 4)$ to obtain five numbers satisfying the conditions of the problem. Thus we found a one-to-one mapping between the set of five numbers satisfying the given conditions and the set of five distinct numbers from the first fourteen positive integers. Therefore the answer is $\binom{14}{5} = 2002$.

12. [AHSME 1978] In a room containing N people, $N > 3$, at least one person has not shaken hands with everyone else in the room. What is the maximum number of people in the room that could have shaken hands with everyone else?

Solution: Label the people $A_1, A_2, \ldots, A_N$ in such a way that A_1 and A_2 are a pair that did not shake hands with each other. Possibly every other pair of people shook hands, so that only A_1 and A_2 did not shake with everyone else. Therefore, at most $N - 2$ people shook hands with everyone else.

13. [AIME 1998] Find the number of ordered quadruples (x_1, x_2, x_3, x_4) of positive odd integers that satisfy $x_1 + x_2 + x_3 + x_4 = 98$.

Solution: Each x_i can be replaced by $2y_i - 1$, where y_i is a positive integer. Because

$$98 = \sum_{i=1}^{4}(2y_i - 1) = 2\left(\sum_{i=1}^{4} y_i\right) - 4$$

it follows that $51 = \sum_{i=1}^{4} y_i$. Each such quadruple (y_1, y_2, y_3, y_4) corresponds in a one-to-one fashion to a row of 51 ones that has been separated into four groups by the insertion of three zeros. For example, (17, 5, 11, 18) corresponds to

$$11111111111111111011111011111111111011111111111111111111.$$

There are $\binom{50}{3} = 19600$ ways to insert three zeros into the fifty spaces between adjacent ones.

14. [USSR 1968] Finitely many cards are placed in two stacks, with more cards in the left stack than in the right. Each card has one or more distinct names written on it, although different cards may share some names. For each name, we define a *shuffle* by moving every card that has that name written on it to the opposite stack. Prove that it is always possible to end up with more cards in the right stack by picking several distinct names, and doing in turn the shuffle corresponding to each name.

Solution: (By Oaz Nir) We will prove the statement by induction on n, the number of distinct names present. Call the left stack L and the right stack R. For case $n = 1$, one shuffle will do the job.

We now assume that we have proved the statement for n names (for some positive integer n), and consider the case with $n + 1$ names. Call the first n names $a_1, a_2, \ldots, a_n$, and let the new name be a. There are two cases.

- *Case 1:* The number of L cards containing only the name a is less than or equal to the number of R cards containing only the name a. We can ignore the name a and use the induction hypothesis to perform required shuffles using some subsets of the n names $a_1, a_2, \ldots, a_n$ and we will be done: there are now more of these remaining cards in stack R than in stack L, and since there were at least as many "only a" cards in R than in L, the final configuration has more cards in R than in L.

- *Case 2:* The number of L cards containing only the name a is greater than the number of R cards containing only the name a. Then we perform one shuffle with the name a, we end up at the beginning of Case 1 and we are done.

In either case, we can finish our inductive step and our proof is complete.

15. [AIME 1992] For how many pairs of consecutive integers in the set

$$\{1000, 1001, 1002, \ldots, 2000\}$$

is no carrying required when the two integers are added?

Solution: Let n have a decimal representation $1abc$. If one of a, b, or c is 5,6,7, or 8, then there will be carrying when n and $n+1$ are added. If $b = 9$ and $c \neq 9$, or if $a = 9$ and either $b \neq 9$ or $c \neq 9$, there will also be carrying when n and $n+1$ are added.

If n is not one of the integers described above, then n has one of the forms

$$1abc \qquad 1ab9 \qquad 1a99 \qquad 1999,$$

where $a, b, c \in \{0, 1, 2, 3, 4\}$. For such n, no carrying will be needed when n and $n+1$ are added. There are $5^3 + 5^2 + 5 + 1 = 156$ such values of n.

16. [AHSME 1994] Nine chairs in a row are to be occupied by six students and Professors Alpha, Beta, and Gamma. These three professors arrive before the six students and decide to choose their chairs so that each professor will be between two students. In how many ways can Professors Alpha, Beta, and Gamma choose their chairs?

First Solution: The two end chairs must be occupied by students, so the professors have seven middle chairs from which to choose, with no two adjacent. If these chairs are numbered from 2 to 8, the three chairs can be:

$$\begin{array}{lllll} (2,4,6), & (2,4,7), & (2,4,8), & (2,5,7), & (2,5,8) \\ (2,6,8), & (3,5,7), & (3,5,8), & (3,6,8), & (4,6,8). \end{array}$$

Within each triple, the professors can arrange themselves in $3!$ ways, so the total number is $10 \times 6 = 60$.

Second Solution: Imagine the six students standing in a row before they are seated. There are 5 spaces between them, each of which may be occupied by at most one of the 3 professors. Therefore, there are $P(5, 3) = 5 \times 4 \times 3 = 60$ ways the professors can select their places.

17. Prove that among any 16 distinct positive integers not exceeding 100 there are four different ones, a, b, c, d, such that $a + b = c + d$.

Solution: Let $a_1 < a_2 < \cdots < a_{16}$ denote the 16 numbers. Consider the difference of each pair of those integers. There are $\binom{16}{2} = 120$ such pairs.

Let (a_i, a_j) denote a pair of numbers with $a_i > a_j$. If we have two distinct pairs of numbers (a_{i_1}, a_{i_2}) and (a_{i_3}, a_{i_4}) such that $a_{i_1} - a_{i_2} = a_{i_3} - a_{i_4}$, then we get the desired quadruple $(a, b, c, d) = (a_{i_1}, a_{i_4}, a_{i_2}, a_{i_3})$ unless $a_{i_2} = a_{i_3}$. We say a is *bad* for the pair of pairs (a_{i_1}, a) and (a, a_{i_2}) if $a_{i_1} - a = a - a_{i_2}$ (or $2a = a_{i_1} + a_{i_2}$). Note that we are done if a number a is bad for two pairs of pairs of numbers. Indeed, if a is bad for (a_{i_1}, a), (a, a_{i_2}) and (a_{i_3}, a), (a, a_{i_4}), then $a_{i_1} + a_{i_2} = 2a = a_{i_3} + a_{i_4}$.

Finally, we assume that each a_i is bad for at most one pair of pairs of numbers. For each such pair of pairs of numbers, we take one pair of numbers out of consideration. Hence there are no bad numbers anymore. Then we still have at least $120 - 16 = 104$ pairs of numbers left. The difference of the numbers in each remaining pair ranges from 1 to 99. By the Pigeonhole Principle, some of these differences have the same value. Assume that $a_{i_1} - a_{i_2} = a_{i_3} - a_{i_4}$, then $(a_{i_1}, a_{i_4}, a_{i_2}, a_{i_3})$ satisfies the conditions of the problem.

18. [AHSME 1989] A child has a set of 96 distinct blocks. Each block is of one of 2 materials *(plastic, wood)*, 3 sizes *(small, medium, large)*, 4 colors *(blue, green, red, yellow)*, and 4 shapes *(circle, hexagon, square, triangle)*. How many blocks in the set are different from the "*plastic medium red circle*" in exactly two ways? (The "*wood medium red square*" is such a block.)

Solution: For a block to differ from the given block, there is only 1 choice for a different material, 2 choices for a different size, 3 choices for a different color, and 3 choices for a different shape. There are $\binom{4}{2} = 6$ ways a block can differ from the block in exactly two ways:

(1) *Material and size:* $1 \cdot 2 = 2$ differing blocks.

(2) *Material and color:* $1 \cdot 3 = 3$ differing blocks.

(3) *Material and shape:* $1 \cdot 3 = 3$ differing blocks.

(4) *Size and color:* $2 \cdot 3 = 6$ differing blocks.

(5) *Size and shape:* $2 \cdot 3 = 6$ differing blocks.
(6) *Color and shape:* $3 \cdot 3 = 9$ differing blocks.

Thus, $2 + 3 + 3 + 6 + 6 + 9 = 29$ blocks differ from the given block in exactly two ways.

19. [AHSME 1998] Call a 7-digit telephone number $d_1d_2d_3 - d_4d_5d_6d_7$ *memorable* if the prefix sequence $d_1d_2d_3$ is exactly the same as either of the sequences $d_4d_5d_6$ or $d_5d_6d_7$ (possibly both). Assuming that each d_i can be any of the ten decimal digits $0, 1, 2, \ldots, 9$, find the number of different memorable telephone numbers.

First Solution: There are 10,000 ways to write the last four digits $d_4d_5d_6d_7$, and among these there are $10000 - 10 = 9990$ for which not all the digits are the same. For each of these, there are exactly two ways to adjoin the three digits $d_1d_2d_3$ to obtain a memorable number. There are ten memorable numbers for which the last four digits are the same, for a total of $2 \cdot 9990 + 10 = 19990$.

Second Solution: Let A denote the set of telephone numbers for which $d_1d_2d_3$ is the same as $d_4d_5d_6$ and let B be the set of telephone numbers for which $d_1d_2d_3$ coincides with $d_5d_6d_7$. A telephone number $d_1d_2d_3-d_4d_5d_6d_7$ belongs to $A \cap B$ if and only if $d_1 = d_4 = d_5 = d_2 = d_6 = d_3 = d_7$. Hence, $n(A \cap B) = 10$. Thus, by the Inclusion-Exclusion Principle,

$$\begin{aligned} n(A \cup B) &= n(A) + n(B) - n(A \cap B) \\ &= 10^3 \cdot 1 \cdot 10 + 10^3 \cdot 10 \cdot 1 - 10 = 19990. \end{aligned}$$

20. [AIME 1996] Two of the squares of a 7×7 checkerboard are painted yellow, and the rest are painted green. Two color schemes are equivalent if one can be obtained from the other by applying a rotation in the plane of the board. How many inequivalent color schemes are possible?

Solution: There are $\binom{49}{2} = 1176$ ways to select the positions of the yellow squares. Because quarter-turns can be applied to the board, however, there are fewer than 1176 inequivalent color schemes. Color schemes in which the two yellow squares are *not* diametrically opposed appear in four equivalent forms. Color schemes in which the two yellow squares *are* diametrically opposed appear in two equivalent forms, and there are $\frac{49-1}{2} = 24$ such pairs of yellow squares. Thus the number of inequivalent color schemes is

$$\frac{1176 - 24}{4} + \frac{24}{2} = 300.$$

21. [ARML 1999] In how many ways can one arrange the numbers 21, 31, 41, 51, 61, 71, and 81 such that the sum of every four consecutive numbers is divisible by 3?

Solution: Since we only need to consider the problem modulo 3, we rewrite the numbers 21, 31, 41, 51, 61, 71, 81 as 0, 1, 2, 0, 1, 2, 0. Suppose that $a_1, a_2, \ldots, a_7$ is a required arrangement. We observe that $0 \equiv (a_1 + a_2 + a_3 + a_4) + (a_4 + a_5 + a_6 + a_7) \equiv (a_1 + a_2 + \cdots + a_7) + a_4 \equiv (0+1+2+0+1+2+0) + a_4 \equiv a_4 \pmod 3$. Thus a_1, a_2, a_3 must be an arrangement of 0, 1, 2 as $a_1 + a_2 + a_3 \equiv a_1 + a_2 + a_3 + a_4 \equiv 0 \pmod 3$. Since $a_1 + a_2 + a_3 + a_4 \equiv a_2 + a_3 + a_4 + a_5 \equiv 0 \pmod 3$, we have $a_1 \equiv a_5 \pmod 3$. Similarly, we can prove that the order of a_5, a_6, a_7 is uniquely determined by a_1, a_2, a_3. Thus we have $3 \times 2^3 \times 3! = 144$ arrangements.

22. [AIME 1993] Let S be a set with six elements. In how many different ways can one select two not necessarily distinct subsets of S so that the union of the two subsets is S? The order of the selection does not matter; for example the pair of subsets $\{a, c\}$, $\{b, c, d, e, f\}$ represents the same selection as the pair $\{b, c, d, e, f\}$, $\{a, c\}$.

Solution: In order that $A \cup B = S$, for each element s of S exactly one of the following three statements is true:

$$s \in A \text{ and } s \notin B \qquad s \notin A \text{ and } s \in B \qquad s \in A \text{ and } s \in B.$$

Hence if S has n elements, there are 3^n ways to choose the sets A and B. Except for pairs with $A = B$, this total counts each pair of sets twice. Since $A \cup B = S$ with $A = B$ occurs if and only if $A = B = S$, the number of pairs of subsets of S whose union is S is

$$\frac{3^n - 1}{2} + 1,$$

which is 365 when $n = 6$.

23. [AIME 2001] A set of positive numbers has the *triangle property* if it has three distinct elements that are the lengths of the sides of a triangle whose area is positive. Consider sets $\{4, 5, 6, \ldots, n\}$ of consecutive positive integers, all of whose ten-element subsets have the triangle property. What is the largest possible value of n?

Solution: The set $\{4, 5, 9, 14, 23, 37, 60, 97, 157, 254\}$ is a ten-element subset of $\{4, 5, 6, \ldots, 254\}$ that does not have the triangle property. Let N be the smallest integer for which $\{4, 5, 6, \ldots, N\}$ has a ten-element subset that lacks the triangle property. Let $\{a_1, a_2, a_3, \ldots, a_{10}\}$ be such a subset, with $a_1 < a_2 < a_3 < \cdots < a_{10}$. Because none of its three-element subsets define triangles, the following must be true:

$$\begin{aligned} N &\geq a_{10} \geq a_9 + a_8 \geq (a_8 + a_7) + a_8 \\ &= 2a_8 + a_7 \geq 2(a_7 + a_6) + a_7 = 3a_7 + 2a_6 \\ &\geq 3(a_6 + a_5) + 2a_6 = 5a_6 + 3a_5 \geq 8a_5 + 5a_4 \\ &\geq 13a_4 + 8a_3 \geq 21a_3 + 13a_2 \geq 34a_2 + 21a_1 \\ &\geq 34 \cdot 5 + 21 \cdot 4 = 254 \end{aligned}$$

Thus the largest possible value of n is $N - 1 = 253$. This is yet another application of the **Fibonacci sequence.**

24. [MOSP 1997] Let A and B be disjoint sets whose union is the set of natural numbers. Show that for every natural number n there exist distinct $a, b > n$ such that

$$\{a, b, a + b\} \subseteq A \quad \text{or} \quad \{a, b, a + b\} \subseteq B.$$

Solution: We shall construct numbers $a, b > n$ such that $a + b$ is in the same set as a and b. First assume that $|A|$ is finite and that m is its largest element. Then $n + 1, n + 2$, and $2n + 3 = (n + 1) + (n + 2)$ are all in B for all $n \geq m$. Consequently, we assume that both A and B are infinite sets.

We approach indirectly. Assume that there is a positive integer n such that for any $a, b > n, \{a, b, a + b\} \not\subset A$ and $\{a, b, a + b\} \not\subset B$. We now choose x, y, and z in A such that $x > y > z > n$ and $y - z > n$. This is possible since A is infinite and thus unbounded. Then $\{x + y, y + z, z + x\} \subset B$. But then $y - z$ has no place to go. Hence our assumption was wrong and we are done.

25. [AIME 1986] The increasing sequence $1, 3, 4, 9, 10, 12, 13, \ldots$ consists of all those positive integers which are powers of 3 or sums of distinct powers of 3. Find the 100þ term of this sequence (where 1 is the 1st term, 3 is the 2nd term, and so on).

First Solution: If we use only the first six non-negative integral powers of 3, namely $1, 3, 9, 27, 81$ and 243, then we can form only 63 terms, since

$$\binom{6}{1} + \binom{6}{2} + \cdots + \binom{6}{6} = 2^6 - 1 = 63.$$

Consequently, the next highest power of 3, namely 729, is also needed.

After the first 63 terms of the sequence the next largest ones will have 729 but not 243 as a summand. There are 32 of these, since $\binom{5}{0}+\binom{5}{1}+\cdots+\binom{5}{5}=32$, bringing the total number of terms to 95. Since we need the 100þ term, we must next include 243 and omit 81. Doing so, we find that the 96þ, 97þ, ..., 100þ terms are: 729+243, 729+243+1, 729+243+3, 729+243+3+1, and $729+243+9=981$.

Second Solution: Note that a positive integer is a term of this sequence if and only if its base 3 representation consists only of 0's and 1's. Therefore, we can set up a one-to-one correspondence between the positive integers and the terms of this sequence by representing both with binary digits (0's and 1's), first in base 2 and then in base 3:

$$\begin{aligned}
1 = 1_{(2)} &\iff 1_{(3)} = 1\\
2 = 10_{(2)} &\iff 10_{(3)} = 3\\
3 = 11_{(2)} &\iff 11_{(3)} = 4\\
4 = 100_{(2)} &\iff 100_{(3)} = 9\\
5 = 101_{(2)} &\iff 101_{(3)} = 10\\
&\vdots
\end{aligned}$$

This is a correspondence between the two sequences in the order given, that is, the *k*þ positive integer is made to correspond to the *k*þ sum (in increasing order) of distinct powers of 3. This is because when the binary numbers are written in increasing order, they are still in increasing order when interpreted in any other base. (If you can explain why this is true when interpreted in base 10, you should be able to explain it in base 3 as well.)

Therefore, to find the 100þ term of the sequence, we need only look at the 100þ line of the above correspondence:

$$100 = 1100100_{(2)} \iff 1100100_{(3)} = 981.$$

26. [AIME 1997] Every card in a deck has a picture of one shape—circle, square, or triangle, which is painted in one of three colors — red, blue, or green. Furthermore, each color is applied in one of three shades — light, medium, or dark. The deck has 27 cards, with every shape-color-shade combination represented. A set of three cards from the deck is called *complementary* if all of the following statements are true:

    (a) Either each of the three cards has a different shape or all three of the cards have the same shape.

(b) Either each of the three cards has a different color or all three of the cards have the same color.

(c) Either each of the three cards has a different shade or all three of the cards have the same shade.

How many different complementary three-card sets are there?

Solution: Consider any pair of cards from the deck. We show that there is exactly one card that can be added to this pair to make a complementary set. If the cards in the pair have the same shape, then the third card must also have this shape, while if the cards have different shapes, then the third card must have the one shape that differs from them. In either case, the shape on the third card is uniquely determined. Similar reasoning shows that the color and the shade on the third card are also uniquely determined. The third card, determined by the first two, is never one of the first two cards. Thus we can count the number of complementary sets by counting the number of pairs of cards and then dividing by 3, because each complementary set is counted three times by this procedure. The number of complementary sets is

$$\frac{1}{3}\binom{27}{2} = \frac{1}{3} \cdot \frac{27 \cdot 26}{2} = 117.$$

27. [China 1990] At a math camp, every m students share exactly one common friend, $m \geq 3$. (If A is a friend of B, then B is a friend of A. Also, A is not his own friend.) Suppose person P has the largest number of friends. Determine what that number is.

First Solution: First note that every student has a friend. Assume that students $A_1, A_2, \ldots, A_k$ are friends of each other, where k is a positive integer, $2 \leq k \leq m$. Then there is a student A_{k+1} who is a common friend to all of the students A_i, $1 \leq i \leq k$. Thus, we can start with a pair of students A_1, A_2 who are friends, and keep adding a student until we obtain $m + 1$ students $A_1, A_2, \ldots, A_{m+1}$ who are friends of each other.

We claim that there are no students other than $A_1, A_2, \ldots, A_{m+1}$ in the camp. For the sake of contradiction, assume there is another student B in the camp. Then B must have a friend. We consider the following situations.

- *Case 1:* If B has at least two friends among the students $A_1, A_2, \ldots, A_{m+1}$, we assume without loss of generality that A_1 and A_2 are friends of B. Then the m students $B, A_3, A_4, \ldots, A_{m+1}$ have two common friends A_1 and A_2, which contradicts the conditions of the problem.

- *Case 2:* If B has no more than one friend among the students A_1, $A_2, \ldots, A_{m+1}$ in the camp, we assume without loss of generality that $A_2, A_3, \ldots, A_{m+1}$ are not friends of B. Then the m students $B, A_1, A_3, \ldots, A_m$ have a common friend C, $C \neq A_i$ for $1 \leq i \leq m+1$. But since $m \geq 3$, student C has at least 2 friends among the students $A_1, A_2, \ldots A_{m+1}$. But this impossible by our argument in Case 1.

Overall, we showed that this camp only has the $m+1$ students $A_1, A_2, \ldots$, A_{m+1} and they are all friends of each other. Hence the desired number is m.

Second Solution: First we observe that P must have at least m friends, since for any set of m students, their common friend has at least m friends (namely, those m students). Now we prove that P cannot have more than m friends. Assume the contrary. Let S be the set of P's friends, and let $n = |S|$. We have by assumption $n \geq m+1$. We claim that for each $(m-1)$-element subset S' of S, there exists a unique person $Q_{S'} \in S$ who is a common friend of all the people in S'.

Consider any such subset S'. Adding P to this set gives a set of size m, and thus by the given set there exists a unique person Q who is a friend of P and all of the members of S'. We claim that this Q is the $Q_{S'}$ that we want. Indeed, $Q \in S$ since by definition, S is the set of all friends of P.

Now we claim that for any two distinct $(m-1)$-element subsets S_1 and S_2 of S, $Q_{S_1} \neq Q_{S_2}$. Assume for a contradiction that this is not the case, that is, there exist $S_1, S_2 \subset S$ with $Q_{S_1} = Q_{S_2}$. Take any m-element subset of $S_1 \cup S_2$. Then the people in this set have two mutual friends, Q_{S_1} and P, contradicting the given.

It follows that each $(m-1)$-element subset S' corresponds to a different person $Q_{S'}$. Now, the number of $m-1$-element subsets of S is

$$\binom{n}{m-1} \geq \binom{n}{2} > n,$$

since $n \geq m+1$ and $m \geq 3$. But $n = |S|$, so two of the Q's must be the same, a contradiction.

28. [AHSME 1989] Suppose that 7 boys and 13 girls line up in a row. Let S be the number of places in the row where a boy and a girl are standing next to each other. For example, for the row $GBBGGGBGBGGGBGBGGBGG$ we have $S = 12$. Find the average value of S (if all possible orders of these 20 people are considered).

First Solution: Suppose that John and Carol are two of the people. For $i = 1, 2, \ldots, 19$, let J_i and C_i be the numbers of orderings (out of all 20!) in which the iþ and $(i+1)^{\text{st}}$ persons are John and Carol, or Carol and John, respectively. Then $J_i = C_i = 18!$ is the number of orderings of the remaining persons.

For $i = 1, 2, \ldots, 19$, let N_i be the number of times a boy-girl or girl-boy pair occupies positions i and $i+1$. Since there are 7 boys and 13 girls, $N_i = 7 \cdot 13 \cdot (J_i + C_i)$. Thus the average value of S is

$$\frac{N_1 + N_2 + N_3 + \ldots + N_{19}}{20!} = \frac{19[7 \cdot 13 \cdot (18! + 18!)]}{20!} = \frac{91}{10}.$$

Second Solution: In general, suppose there are k boys and $n-k$ girls. For $i = 1, 2, \ldots, n-1$, let A_i be the probability that there is a boy-girl pair in positions $(i, i+1)$ in the line. Since there is either 0 or 1 pair in $(i, i+1)$, A_i is also the expected number of pairs in these positions. By symmetry, all A_i's are the same (or note that the argument below is independent of i). Thus, the answer is $(n-1)A_i$.

We may consider the boys indistinguishable and likewise the girls. (*Why?*) Then an order is just a sequence of k **B**s and $n-k$ **G**s. To have a pair at $(i, i+1)$ we must have **BG** or **GB** in those positions, and the remaining $n-2$ positions must have $k-1$ boys and $n-k-1$ girls. Thus there are $2\binom{n-2}{k-1}$ sequences with a pair at $(i, i+1)$. Since there are $\binom{n}{k}$ sequences, the answer is

$$(n-1)A_i = \frac{(n-1)2\binom{n-2}{k-1}}{\binom{n}{k}} = \frac{2k(n-k)}{n}.$$

Thus, when $n = 20$ and $k = 7$, the answer is $(2 \cdot 7 \cdot 13)/20 = 91/10$.

29. [AIME 1996] A bored student walks down a hall that contains a row of closed lockers, numbered 1 to 1024. He opens the locker numbered 1, and then alternates between skipping and opening each closed locker thereafter. When he reaches the end of the hall, the student turns around and starts back. He opens the first closed locker he encounters, and then alternates between skipping and opening each closed locker thereafter. The student continues wandering back and forth in this manner until every locker is open. What is the number of the last locker he opens?

First Solution: Suppose that there are 2^k lockers in the row, and let L_k be the number of the last locker opened. After the student makes his first pass along the row, there are 2^{k-1} closed lockers left. These closed lockers all have even numbers and are in descending order from where the student is standing. Now, renumber the closed lockers from 1 to 2^{k-1}, starting from the end where the student is standing. Notice that the locker originally numbered n (where n is even) is now numbered $2^{k-1}+1-\frac{n}{2}$. Thus, because L_{k-1} is the number of the last locker opened with this new numbering, we have

$$L_{k-1} = 2^{k-1} + 1 - \frac{L_k}{2}.$$

Solving for L_k we find

$$L_k = 2^k + 2 - 2L_{k-1}.$$

Iterate this recursion once to obtain

$$L_k = 2^k + 2 - 2(2^{k-1} + 2 - 2L_{k-2}) = 4L_{k-2} - 2. \tag{1}$$

When there are $1024 = 2^{10}$ lockers to start with, the last locker to be opened is numbered L_{10}. Apply (1) repeatedly to $L_0 = 1$ to find that $L_2 = 4L_0 - 2 = 2$, $L_4 = 6$, $L_6 = 22$, $L_8 = 86$, and $L_{10} = 342$.

Second Solution: Follow the given solution to the recursion (1), which can be written in the form

$$L_k - \frac{2}{3} = 4\left(L_{k-2} - \frac{2}{3}\right).$$

Because $L_0 = 1$ and $L_1 = 2$, it follows that

$$L_k - \frac{2}{3} = \begin{cases} \left(1 - \frac{2}{3}\right)4^{\frac{k}{2}} & \text{if } k \text{ is even,} \\ \left(2 - \frac{2}{3}\right)4^{\frac{k-1}{2}} & \text{if } k \text{ is odd .} \end{cases}$$

These formulas may be combined to yield

$$L_k = \frac{1}{3}\left(4^{\left\lfloor \frac{k+1}{2} \right\rfloor} + 2\right)$$

for all nonnegative k. In particular, $L_{10} = 342$.

Note: How would the solution change if there were 1000 lockers in the hall?

30. [AIME 1995] Let $n = 2^{31}3^{19}$. How many positive integer divisors of n^2 are less than n but do not divide n?

First Solution: Let $n = p^r q^s$, where p and q are distinct primes. Then $n^2 = p^{2r} q^{2s}$, so n^2 has

$$(2r+1)(2s+1)$$

factors. For each factor less than n, there is a corresponding factor greater than n. By excluding the factor n, we see that there must be

$$\frac{(2r+1)(2s+1)-1}{2} = 2rs + r + s$$

factors of n^2 that are less than n. Because n has $(r+1)(s+1)$ factors (including n itself), and because every factor of n is also a factor of n^2, there are

$$2rs + r + s - [(r+1)(s+1) - 1] = rs$$

factors of n^2 that are less than n but not factors of n. When $r = 31$ and $s = 19$, there are $rs = 589$ such factors.

Second Solution: (By Chengde Feng) A positive integer divisor d of n^2 is less than n but does not divide n if and only if

$$d = \begin{cases} 2^{31+a} 3^{19-b} & \text{if } 2^a < 3^b, \\ 2^{31-a} 3^{19+b} & \text{if } 2^a > 3^b, \end{cases}$$

where a and b are integers such that $1 \le a \le 31$ and $1 \le b \le 19$. Since $2^a \ne 3^b$ for positive integers a and b, there are $19 \times 31 = 589$ such divisors.

31. [China 1990] In an arena, each row has 199 seats. One day, 1990 students are coming to attend a soccer match. It is only known that at most 39 students are from the same school. If students from the same school must sit in the same row, determine the minimum number of rows that must be reserved for these students.

Solution: Since 199 is a prime, we consider 200. Its largest divisor not exceeding 39 is 25. Note that $1990 = 79 \times 25 + 15$. If 79 schools send 25 students each and one school sends 15 students, it will take at least $\lceil 79/\lfloor 199/25 \rfloor \rceil = 12$ rows to seat all the students.

We now prove that 12 rows are enough. Start seating the students school by school and row by row, filling all the seats of the first 10 rows, even if students from some schools are split between two rows. This can happen to at most 9 schools. Remove the students from those schools and pack them into two rows. This is possible since each row can hold students from at least 5 schools as $5 \times 39 = 195 < 199$.

Note: Interested readers might want to solve this dual version:

In an arena, there are 11 rows of seats and each row has 199 seats. One day, n students are coming to attend a basketball match. It is only known that at most 39 students are from the same school. If students from the same school must sit in the same row, determine the maximum number of students such that all the students will be seated.

32. [AIME 1990] Let $T = \{9^k \mid k \text{ is an integer}, 0 \le k \le 4000\}$. Given that 9^{4000} has 3817 digits and that its first (leftmost) digit is 9, how many elements of T have 9 as their leftmost digit?

Solution: Note that 9^k has one more digit than 9^{k-1}, except in the case when 9^k starts with a 9. In the latter case, long division shows that 9^{k-1} starts with a 1 and has the same number of digits as 9^k. Therefore, when the powers of 9 from 9^0 to 9^{4000} are computed there are 3816 increases in the number of digits. Thus there must be $4000 - 3816 = 184$ instances when computing 9^k from 9^{k-1} $(1 \le k \le 4000)$ does not increase the number of digits. Since $9^0 = 1$ does not have leading digit 9 we can conclude that 9^k $(1 \le k \le 4000)$ has a leading digit of 9 exactly when there is no increase in the number of digits in computing 9^k from 9^{k-1}. It follows that 184 of the numbers must start with the digit 9.

Note: We did not need to know that the leading digit of 9^{4000} is 9, but it was important to note that the leading digit of 9^0 is *not* 9.

33. [USAMO 1999 submission, Jim Propp] For what values of $n \ge 1$ do there exist a number m that can be written in the form $a_1 + \cdots + a_n$ (with $a_1 \in \{1\}, a_2 \in \{1, 2\}, \ldots, a_n \in \{1, \ldots, n\}$) in $(n-1)!$ or more ways?

First Solution: Note that for $n = 1, 2, 3, 4$, we may choose $m = 1, 3, 5, 7$, respectively.

Note that each of the ways to write the number m in the form $a_1 + \cdots + a_n$ (with $a_1 \in \{1\}, a_2 \in \{1, 2\}, \ldots, a_n \in \{1, \ldots, n\}$) requires a different ordered $(n-1)$-tuple $(a_1, a_2, \ldots, a_{n-1})$. Furthermore, there are only $(n-1)!$ such $(n-1)$-tuples, so each of those must work for m; i.e., we must have

$$2n - 1 = \underbrace{1 + 1 + \cdots + 1}_{n-1 \text{ 1's}} + n \ge m$$

or else there would be no valid expression for m with $a_1 = a_2 = \cdots = 1$, and also

$$m \geq 1+2+\cdots+(n-1)+1 = \frac{n(n-1)}{2}+1,$$

or else there would be no valid expression for m with $a_1 = 1$, $a_2 = 2, \ldots, a_{n-1} = n-1$. Combining the two inequalities above, we have

$$2(n-1) \geq \frac{n(n-1)}{2},$$

or $n \leq 4$.

Hence $n = 1, 2, 3, 4$ are the only n satisfying the conditions of the problem.

Second Solution: (David Vincent) For each n, define the polynomial

$$f_n(x) = x(x+x^2)\cdots(x+x^2+\cdots+x^n).$$

It is clear that $f_n(x)$ is a $1+2+\cdots+n = \frac{n(n+1)}{2}$ degree polynomial. We can write

$$f_n(x) = f_{n,1}x + f_{n,2}x^2 + \cdots + f_{n,\frac{n(n+1)}{2}}x^{\frac{n(n+1)}{2}}.$$

Then the coefficient of the term x^m in $f_n(x)$, $[x^m](f_n(x)) = f_{n,m}$, is equal to the number of ways that m can be written in the form $a_1 + \cdots + a_n$ with $a_1 \in \{1\}, a_2 \in \{1, 2\}, \ldots, a_n \in \{1, \ldots, n\}$. For convenience, we may extend this definition to the other powers of x by letting $f_{n,m} = 0$ for all m not yet mentioned.

We have

$$\begin{aligned}
f_1(x) &= x, \\
f_2(x) &= x^2 + x^3, \\
f_3(x) &= x^3 + 2x^4 + 2x^5 + x^6, \\
f_4(x) &= x^4 + 3x^5 + 5x^6 + 6x^7 + 5x^8 + 3x^9 + x^{10}.
\end{aligned}$$

It follows that for $n = 1, 2, 3, 4$, $m = 1$; $m = 2$ or $m = 3$; $m = 4$ or $m = 5$; $m = 7$, work, respectively.

It is not difficult to see that there are $(1+2\cdots+n) - n + 1 = \frac{n(n-1)}{2} + 1$ terms in $f_n(x)$. For $n \geq 5$, it is also not difficult to see that $f_{n-1}(x)$ has

$$\frac{(n-1)(n-2)}{2} + 1 > n$$

terms. Since $f_n(x) = f_{n-1}(x)(x + x^2 + \cdots x^n)$, for positive integers m,

$$
\begin{aligned}
&[x^m](f_n(x)) \\
&= f_{n-1,m-1} + f_{n-1,m-2} + \cdots + f_{n-1,m-n} \\
&< \sum_{i=1}^{\frac{n(n-1)}{2}} f_{n-1,i} = f_{n-1}(1) = (n-1)!.
\end{aligned}
$$

34. [AIME 1986] Let the sum of a set of numbers be the sum of its elements. Let S be a set of positive integers, none greater than 15. Suppose no two disjoint subsets of S have the same sum. What is the largest sum a set S with these properties can have?

Solution: First we show that S contains at most 5 elements. Suppose otherwise. Then S has at least $\binom{6}{1} + \binom{6}{2} + \binom{6}{3} + \binom{6}{4}$ or 56 subsets of 4 or fewer members. The sum of each of these subsets is at most 54 (since $15 + 14 + 13 + 12 = 54$); hence, by the Pigeonhole Principle, at least two of these sums are equal. If the subsets are disjoint, we are done; if not, then the removal of the common element(s) yields the desired contradiction.

It is not difficult to show that the set $S' = \{15, 14, 13, 11, 8\}$, on the other hand, satisfies the conditions of the problem. The sum of S' is 61. Hence the set S we seek is a 5-element set with a sum of at least 61. Let $S = \{a, b, c, d, e\}$ with $a < b < c < d < e$, and let s denote the sum of S. Then it is clear that $d + e \leq 29$ and $c \leq 13$. Since there are $\binom{5}{2}$ 2-element subsets of S, $a+b \leq d+e-10+1 \leq 20$. Hence $s \leq 20+13+29 = 62$. If $c \leq 12$, then $S \leq 61$; if $c = 13$, then $d = 14$ and $e = 15$. Then $s \leq a + b + 42$. Since $12+15 = 13+14$, $b \leq 11$. If $b \leq 10$, then $a+b \leq 19$ and $s \leq 61$; if $b = 11$, then $a \leq 8$ as $10 + 15 = 11 + 14$ and $9 + 15 = 11 + 13$, implying that $s \leq 8+11+42 = 61$. In all cases, $s \leq 61$. It follows that the maximum we seek is 61.

35. [China 1994, Zonghu Qiu] There are at least four candy bars in n ($n \geq 4$) boxes. Each time, Mr. Fat is allowed to pick two boxes, take one candy bar from each of the two boxes, and put those candy bars into a third box. Determine if it is always possible to put all the candy bars into one box.

Solution: It is always possible to put all the candy bars into one box. We will prove our statement by induction on m, the number of candy bars.

For the base case $m = 4$, there are at most 4 nonempty boxes. We disregard all the other empty boxes and consider all the possible initial distributions:

(1) $(1, 1, 1, 1)$ (2) $(1, 2, 1, 0)$ (3) $(2, 2, 0, 0)$ (4) $(1, 3, 0, 0)$.
For distribution (1), we proceed as follows:

$$(1, 1, 1, 1) \to (3, 1, 0, 0) \to (2, 0, 2, 0) \to (1, 0, 1, 2) \to (0, 0, 0, 4).$$

It is easy to see that all the other initial distributions are covered in the above sequence of operations. Thus the base case is proved.

Now we assume that the statement is true for some positive integer $m \geq 4$. If we are given $m + 1$ candy bars, we mark one of them and called it *special*. We first ignore the special candy bar and consider only the other m candy bars. By the induction hypothesis, we can put all m candy bars into one box. If this box also contains the special piece, we are done. If not, we pick two empty boxes and proceed as follows:

$$\begin{aligned} &(1, m, 0, 0) \to (0, m - 1, 2, 0) \to (0, m - 2, 1, 2) \\ \to\ &(2, m - 3, 0, 2) \to (1, m - 1, 0, 1) \to (0, m + 1, 0, 0). \end{aligned}$$

Now all the candy bars are in one box and our induction is complete.

36. Determine, with proof, if it is possible to arrange $1, 2, \ldots, 1000$ in a row such that the average of any pair of distinct numbers is not located in between the two numbers.

Solution: We claim that it is possible to arrange $1, 2, \ldots, n$ in a row such that the average of any pair of distinct numbers is not located in between the two numbers.

We first prove that this is true for $n = 2^m$ for all positive integers m. We induct on m. The base $m = 1$ is trivial.

Now we assume that we can arrange $1, 2, \ldots, 2^m$, for some positive integer m, in a row $(a_1, a_2, \ldots, a_{2^m})$ such that the average of any pair of distinct numbers is not located in between the two numbers. It is not difficult to see that

$$\begin{aligned} &(b_1, b_2, \ldots, b_{2^{m+1}}) \\ =\ &(2a_1 - 1, 2a_2 - 1, \ldots, 2a_{2^m} - 1, 2a_1, 2a_2, \ldots, 2a_{2^m}) \end{aligned}$$

is an arrangement of the numbers $1, 2, \ldots, 2^{m+1}$ satisfying the conditions of the problem. Indeed, the average of a pair of numbers b_i and b_j with either $1 \leq i < j \leq 2^m$ or $2^m + 1 \leq i < j \leq 2^{m+1}$ is not located between the two numbers by the induction hypothesis, and the average of a pair of numbers b_i and b_j with $1 \leq i \leq 2^m < j \leq 2^{m+1}$ is not an integer. Our induction is thus complete.

For a positive integer n that is not a power of 2, we can always find a positive integer m such that $n < 2^m$. We first arrange the numbers $1, 2, \ldots, 2^m$ in the desired fashion and then delete all the numbers that are larger than n to obtain an arrangement of the numbers $1, 2, \ldots, n$ satisfying the conditions of the problem.

37. Let $A_1A_2 \ldots A_{12}$ be a regular dodecagon with O as its center. Triangular regions OA_iA_{i+1}, $1 \le i \le 12$ (and $A_{13} = A_1$) are to be colored red, blue, green, or yellow such that adjacent regions are colored in different colors. In how many ways can this be done?

Solution: We will find a general formula. Let $A_1A_2 \ldots A_n$ $(n \ge 3)$ be a regular n-sided polygon with O as its center. Triangular regions OA_iA_{i+1}, $1 \le i \le n$ (and $A_{n+1} = A_1$) are to be colored in one of the k $(k \ge 3)$ colors such that adjacent regions are colored in different colors. Let $p_{n,k}$ denote the number of such colorings. We want to find $p_{12,4}$.

There are k ways to color the region OA_1A_2, and then $k-1$ ways to color regions OA_2A_3, OA_3A_4, and so on. We have to be careful about the coloring of the region OA_nA_1. It is possible that it has the same color as that of region OA_1A_2. But then, we simply end up with a legal coloring for $n-1$ regions by viewing region OA_nA_2 as one region. This is a clear bijection between this special kind of illegal colorings of n regions to legal colorings of $n-1$ regions. Hence $p_{n,k} = k(k-1)^{n-1} - p_{n-1,k}$. Note that $p_{3,k} = k(k-1)(k-2)$. It follows that

$$\begin{aligned}
p_{n,k} &= k(k-1)^{n-1} - k(k-1)^{n-2} + k(k-1)^{n-3} - \cdots \\
&\quad + (-1)^{n-4}k(k-1)^3 + (-1)^{n-3}k(k-1)(k-2) \\
&= k \cdot \frac{(k-1)^n + (-1)^{n-4}(k-1)^3}{1 + (k-1)} + (-1)^{n-3}k(k-1)(k-2) \\
&= (k-1)^n + (-1)^n(k-1)^3 + (-1)^{n-1}k(k-1)(k-2) \\
&= (k-1)^n + (-1)^n(k-1)[(k-1)^2 - k(k-2)] \\
&= (k-1)^n + (-1)^n(k-1).
\end{aligned}$$

Hence $p_{12,4} = 3^{12} + 3 = 531,444$ legal ways to color this regular dodecagon.

38. There are $2n$ people at a party. Each person has an even number of friends at the party. (Here friendship is a mutual relationship.) Prove that there are two people who have an even number of common friends at the party.

Solution: Assume for a contradiction that every two of the people at the party share an odd number of friends. Consider any person P. Let A be the set of P's friends, and let B be the set containing everyone else. Observe that since $|A|$ is even and the total number of people at the party is $2n$, $|B|$ is odd. Consider any person Q in B. By definition of B, Q is not a friend of P. By assumption, Q shares an odd number of friends with P, so Q has an odd number of friends in A. Since the total number of friends of Q is even, Q must also have an odd number of friends in B. Now, summing the number of friends in B over all the Q's in B, we should obtain twice the number of friendships among people in B. But the sum is odd, since as noted previously, $|B|$ is odd. This is a contradiction, and hence two of the people at the party must share an even number of common friends.

Note: One can show that for every person P at the party there exists a person Q who has an even number of common friends with P at the party. Indeed, let sets A and B be as in the solution. The set B is nonempty, since $|B|$ is odd. There must be a person Q who has an even number of friends in B. Then Q must also have an even number of friends in A. In order to justify this stronger statement we did not use a proof by contradiction.

39. [AIME 1997] How many different 4×4 arrays whose entries are all $1's$ and $-1's$ have the property that the sum of the entries in each row is 0 and the sum of the entries in each column is 0?

Solution: Each row and each column must contain two $1's$ and two $-1's$, so there are $\binom{4}{2} = 6$ ways to fill the first row. There are also six ways to fill the second row. Of these, one way has four matches with the first row, four ways have two matches with the first row, and one way has no matches with the first row. The first case allows one way to fill the third row, the second case allows two ways to fill the third row, and the third case allows six ways to fill the third row. Once the first three rows are filled, the fourth row can be filled in only one way. Thus there are $6(1 \cdot 1 + 4 \cdot 2 + 1 \cdot 6) = 90$ ways to fill the array to satisfy the conditions.

40. [IMO Shortlist 1996] A square of dimensions $(n-1) \times (n-1)$ is divided into $(n-1)^2$ unit squares in the usual manner. Each of the n^2 vertices of these squares is to be colored red or blue. Find the number of different colorings such that each unit square has exactly two red vertices. (Two coloring schemes are regarded as different if at least one vertex is colored differently in the two schemes.)

Solution: Let the vertices in the bottom row be assigned an arbitrary coloring, and suppose that some two adjacent vertices have the same color. Then it is not difficult to see that the coloring of the remaining vertices are fixed. There are $2^n - 2$ colorings of the bottom row with the property that some two adjacent vertices have the same color (as there are a total of 2^n colorings and 2 ways alternates the coloring of adjacent vertices.

If the vertices of the bottom row are colored alternately, this property must be true for each of the other rows as well. Hence each row can be colored in 2 ways for a total of 2^n ways.

Therefore the answer is $2^n - 2 + 2^n = 2^{n+1} - 2$ ways satisfying the conditions of the problem.

41. Sixty-four balls are separated into several piles. At each step we are allowed to apply the following operation. Pick two piles, say pile $\mathcal{A}$ with p balls and pile $\mathcal{B}$ with q balls and $p \geq q$, and then remove q balls from pile $\mathcal{A}$ and put them in pile $\mathcal{B}$. Prove that it is possible to put all the balls into one pile.

Solution: We use induction to prove that it is possible to put all the n balls into one pile if $n = 2^m$ for some nonnegative integer m. The base cases $m = 0$ and $m = 1$ are trivial.

Now we assume that it is possible to put all the 2^m balls into one pile for some positive positive integer m. We will show that it is possible to put 2^{m+1} balls into one pile. We first note that there are an even number of piles each containing an odd number of balls. We match those piles and apply the operation to each pair. Hence after finitely many operations, each of the piles contains an even number of balls. We then bind each pair of balls in each pile to form a *super ball*. Hence we obtain a certain number of piles of 2^m super balls. By our induction hypothesis, we can put these super balls into one pile. Hence all the 2^{m+1} balls are now in one pile and our induction is complete.

42. [USAMO 1999 submission, Richard Stong] A game of solitaire is played with a finite number of nonnegative integers. On the first move the player designates one integer as *large*, and replaces another integer by any nonnegative integer strictly smaller than the designated large integer. On subsequent steps play is similar, except that integer replaced must be the one designated as large on the previous play. Prove that in some finite number of steps play must end.

Solution: Let the integers at any time be $a_1, a_2, \ldots, a_n$, and let ℓ be the index of the integer chosen as large in the previous step. Define the score of the position to be $S = \sum_{i \neq \ell} a_i$. At any step we will choose a new large integer $a_{\ell'}$ (which currently contributes to S but will not after the move), and we will replace a_ℓ (which currently does not contribute to S) with something smaller than $a_{\ell'}$ (which will contribute to the new S). Thus S is decreased by at least 1 on every move. Since S starts with a finite value and $S \geq 0$, play must stop in a finite number of moves.

43. [USAMO 2000 submission, Cecil Rousseau] Given $S \subseteq \{1, 2, \ldots, n\}$, we are allowed to modify it in any one of the following ways:

    (a) if $1 \notin S$, add the element 1;
    (b) if $n \in S$, delete the element n;
    (c) for $1 \leq r \leq n-1$, if $r \in S$ and $r+1 \notin S$, delete the element r and add the element $r+1$.

    Suppose that it is possible by such modifications to obtain a sequence

    $$\emptyset \to \{1\} \to \{2\} \to \cdots \to \{n\},$$

    starting with $\emptyset$ and ending with $\{n\}$, in which each of the 2^n subsets of $\{1, 2, \ldots, n\}$ appears exactly once. Prove that $n = 2^m - 1$ for some m.

    Solution: Let m be the sum of the set elements. Whenever operation (a) or (c) is performed, m increases by 1, and whenever (b) is performed, m decreases by n. If (b) is performed d times in a sequence of k set modifications that starts and ends with the same set S, then $(k-d) - dn = 0$, that is, $k = d(n+1)$. Since adding $\{n\} \to \emptyset$ to the presumed sequence gives

    $$\emptyset \to \{1\} \to \{2\} \to \cdots \to \{n\} \to \emptyset,$$

    a cycle of length $k = 2^n$, we have $(n+1) \mid 2^n$. Thus n must be of the form $2^m - 1$ for some $m \leq n$.

44. [China 1989, Pingshen Tao] There are 2001 coins on a table. For $i = 1, 2, \ldots, 2001$ in succession, one must turn over exactly i coins. Prove that it is always possible either to make all of the coins face up or to make all of the coins face down, but not both.

    Solution: The statement works for any odd number of coins. We prove our statement by induction on n (n odd) the number of coins. The base case $n = 1$ is trivial.

    Suppose the statement is true for $n = 2k - 1$, for some positive integer k. If we are given $n = 2k + 1$ coins, we consider the following cases.

- *Case 1:* There is a coin C_1 facing up and another coin C_2 facing down. We consider the other $2k-1$ coins first. By our induction, we can turn the coins $1, 2, \ldots, 2k-1$ times in succession so all the $2k-1$ coins are in one direction. Without loss of generality, we assume that all the $2k-1$ coins are facing up. Then we turn all these coins together with C_1 and then turn all the $2k+1$ coins so they will be all facing up.
- *Case 2:* All the coins are in the same direction. We arrange the coins around a circle and number them $1, 2, \ldots, 2k+1$ in clockwise order. We first turn coin 1, then coins 2 and 3, and then 4, 5, and 6, and so on along the circle. Then we make a total of $1+2+\cdots+(2k+1) = (k+1)(2k+1)$ turnings and each coin has been turned $k+1$ times. Since they start in the same direction, they end in the same direction.

From the above argument, we can find a way to make all of the $2k+1$ coins facing in one direction after $2k+1$ operations regardless of the initial configuration. Hence our induction is complete.

Now we prove that it is impossible to achieve both final configurations, with all of the coins facing up or all of the coins facing down. We approach indirectly by assuming there is a initial configuration A such that there are procedures T_1 and T_2 that can make all coins all face up and face down, respectively. Then we can start with all coins facing down, reverse all the steps in T_2 to obtain configuration A, and then we proceed with all the steps in T_1 and end up with all coins facing up. Each coin has been turned an odd number of times. Since there are 2001 coins, the total number of turnings is odd. On the other hand, we made

$$2 \times (1+2+\cdots+2001) = 2001 \times 2002,$$

an even number of turnings. We reach a contradiction. Hence our assumption was wrong and one can only obtain exactly one of the two final configurations.

45. [AIME 1983] For $\{1, 2, \ldots, n\}$ and each of its nonempty subsets a unique *alternating sum* is defined as follows: Arrange the numbers in the subset in decreasing order and then, beginning with the largest, alternately add and subtract successive numbers. (For example, the alternating sum for $\{1, 2, 4, 6, 9\}$ is $9-6+4-2+1=6$ and for $\{5\}$ it is simply 5.) Find the sum of all such alternating sums for $n=7$.

Solution: It is easier, perhaps, to generalize the problem (ever so slightly) by considering the alternating sums for all subsets of $\{1, 2, \ldots, n\}$, that is, including the empty set. To include the empty set without affecting the answer we have only to declare that its alternating sum be 0. The subsets of $\{1, 2, \ldots, n\}$ may be divided into two kinds: those that do not contain n and those that do. Moreover, each subset of the first kind may be paired, in a one-to-one correspondence, with a subset of the second kind as follows:

$$\{a_1, a_2, \ldots, a_i\} \longleftrightarrow \{n, a_1, a_2, \ldots, a_i\}.$$

(For the empty set we have the correspondence $\emptyset \longleftrightarrow \{n\}$.) Then, assuming $n > a_1 > a_2 > \cdots > a_i$, the sum of the alternating sums for each such pair of subsets is given by

$$(a_1 - a_2 + \cdots \pm a_i) + (n - a_1 + a_2 - \cdots \mp a_i) = n.$$

And since there are 2^n subsets of $\{1, 2, \ldots, n\}$ and, consequently, 2^{n-1} such pairs of subsets, the required sum is $n2^{n-1}$. Finally, taking $n = 7$, we obtain 448.

46. [AIME 1992] In a game of *Chomp*, two players alternately take "bites" from a 5-by-7 grid of unit squares. To take a bite, the player chooses one of the remaining squares, then removes ("eats") all squares found in the quadrant defined by the left edge (extended upward) and the lower edge (extended rightward) of the chosen square. For example, the bite determined by the shaded square in the diagram would remove the shaded square and the four squares marked by $\times$.

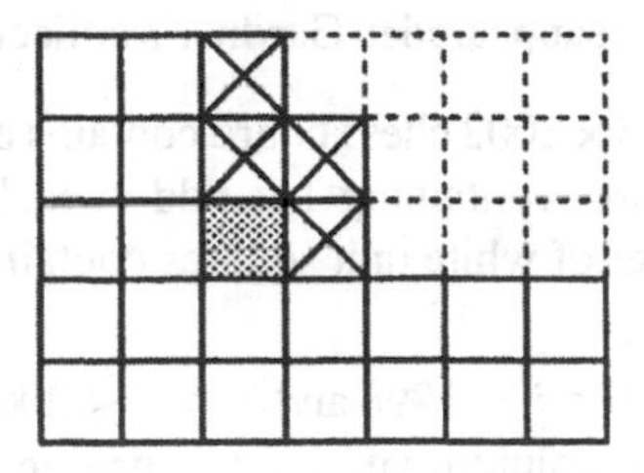

(The squares with two or more dotted edges have been removed from the original board in previous moves.) The object of the game is to make one's opponent take the last bite. The diagram shows one of the many subsets of the set of 35 unit squares that can occur during the game of Chomp. How many different subsets are there in all? Include the full board and the empty board in your count.

Solution: At any stage of the game, the uneaten squares will form columns of nonincreasing heights as we read from left to right.

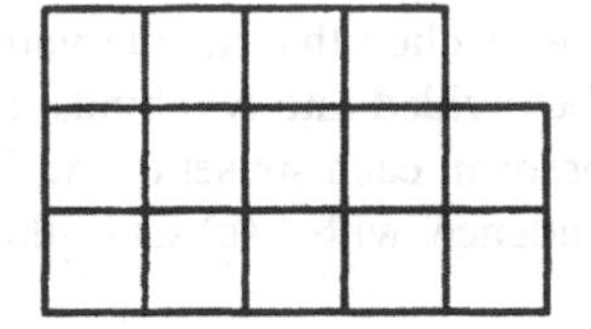

It is not hard to show that this condition is not only necessary, but is also sufficient for a given configuration of squares to occur in a game. (The reader should prove this fact.) Moreover, any such configuration can be completely described by the twelve-step polygonal path that runs from the upper left to the lower right of the original board, forming the boundary between the eaten and uneaten squares. This polygonal boundary can be described by a twelve-letter sequence of $V's$ and $H's$. Such a sequence contains seven $H's$, where each H represents the top of an uneaten column (or bottom of a completely eaten one) and five $V's$, where each V represents a one-unit drop in vertical height in moving from the top of an uneaten column to the top of an adjacent, but shorter column. For example, the state that appears in the diagram accompanying the problem is described by $HHHVHVVHHHVV$, while the sequences $HHHHHHHVVVVV$ and $VVVVVHHHHHHH$ describe the full board and the empty board, respectively. Thus the number of possible subsets is $\binom{12}{7} = 792$.

Note: The game of *Chomp* is due to David Gale, and was introduced (and named) by Martin Gardner in his *Scientific American* column "Mathematical Games". The column reappeared in Gardner's collection *Knotted Doughnuts*.

47. Each square of a 1998×2002 chess board contains either 0 or 1 such that the total number of squares containing 1 is odd in each row and each column. Prove that the number of white unit squares containing 1 is even.

Solution: Let (i, j), $1 \le i \le 1998$ and $1 \le j \le 2002$ denote the unit square in the iþ row and jþ column, and let $a_{i,j}$ denote the number in (i, j). A square (i, j) is white if and only if i and j have the same parity. By the given conditions, the sum

$$R_{\text{odd}} = \sum_{i=1}^{999} \sum_{j=1}^{2002} a_{2i-1,j}$$

is the sum of all the numbers in the 999 odd rows, i.e., R_{odd} is odd as it is the sum of 999 odd numbers. Likewise, sum of all the numbers in even columns

$$C_{\text{even}} = \sum_{j=1}^{1001} \sum_{i=1}^{1998} a_{2j,i}$$

is also odd as it is the sum of 1001 odd numbers. Let B denote the set of all the black squares in the even columns, and let $S(B)$ denote the sum of the numbers in the squares in set B. Note that the numbers in each of the squares in B appears exactly once in the sum R_{odd}. Note also that the numbers in each of the squares in B appear exactly once in the sum C_{even}. Finally, note that each of the numbers in the white square appears exactly once in combined sum $R_{\text{odd}} + C_{\text{even}}$. Thus the total of the numbers of the white unit squares is $R_{\text{odd}} + C_{\text{even}} - 2S(B)$, which is even. Therefore the number of white unit squares containing 1 is even.

48. [AIME 1989] Let S be a subset of $\{1, 2, 3, \ldots, 1989\}$ such that no two members of S differ by 4 or 7. What is the largest number of elements S can have?

Solution: We first show that, given any set of 11 consecutive integers from $\{1, 2, 3, \ldots, 1989\}$, at most five of these 11 can be elements of S. We prove this fact for the set $T = \{1, 2, 3, \ldots, 11\}$, but the same proof works for any set of 11 consecutive integers. Consider the following partition of T, where each subset was formed so that it can contribute at most one element to S:

$$\{1, 5\} \quad \{2, 9\} \quad \{3, 7\} \quad \{4, 11\} \quad \{6, 10\} \quad \{8\}. \tag{2}$$

If it were possible to have 6 elements of T in S, then each of the sets in (2) would have to contribute exactly one element. That this is impossible is shown by the following chain of implications:

$$8 \in S \Rightarrow 1 \notin S \Rightarrow 5 \in S \Rightarrow 9 \notin S \Rightarrow 2 \in S \Rightarrow 6 \notin S \Rightarrow 10 \in S \Rightarrow$$
$$3 \notin S \Rightarrow 7 \in S \Rightarrow 11 \notin S \Rightarrow 4 \in S \Rightarrow 8 \notin S.$$

With the aid of (2), or otherwise, it is easy to find a 5-element subset of T that satisfies the key property of S (i.e., no two numbers differ by 4 or 7). One such set is

$$T' = \{1, 3, 4, 6, 9\}.$$

We also find (perhaps to our surprise) that T' has the remarkable property of allowing for a periodic continuation. That is, if I denotes the set of integers, then

$$S' = \{k + 11n | k \in T' \text{ and } n \in I\}$$

also has the property that no two elements in the set differ by 4 or 7. Moreover, since $1989 = 180 \cdot 11 + 9$, it is clear that S cannot have more than $181 \cdot 5 = 905$ elements. Because the largest element in T' is 9, it follows that the set

$$S = S' \cap \{1, 2, 3, \ldots, 1989\}$$

has 905 elements and hence shows that the upper bound of 905 on the size of the desired set can be attained. This completes the argument.

Note: The reader may wish to find other 5-element subsets of $\{1, 2, 3, \ldots, 11\}$ that exhibit the key property of S. Which of these subsets can be used, as above, to generate a maximal S?

The reader is also encouraged to explore similar problems with other pairs (triples, etc.) of integers in place of 4 and 7, and to find the appropriate motivations for the choice of 11 as the size of the blocks of integers considered in the above solution.

49. [USAMO 2002 submission, Zuming Feng] A class of fifteen boys and fifteen girls is seated around a round table. Their teacher wishes to pair up the students and hand out fifteen tests—one test to each pair.

As the teacher is preparing to select the pairs and hand out the tests, he wonders to himself, "How many seating arrangements would allow me to match up boy/girl pairs sitting next to each other without having to ask any student to change his or her seat?" Answer the teacher's question. (Two seating arrangements are regarded as being the same if one can be obtained from the other by a rotation.)

Solution: We call a pairing *good* if each contains a boy and a girl. It is clear that there are 15! good pairings. For each good pairing, there are $14! \times 2^{15}$ ways to arrange the students around the table. Such a seating arrangement is called a *good working relation*. Hence there are a total of $14! \times 15! \times 2^{15}$ *good working relations*.

We call a seating arrangement *good* if it allows the teacher to match up boy/girl pairs sitting next to each other without having to ask any student to change his or her seat. We want to evaluate x, the number of good arrangements. There are two types of good seating arrangements:

(a) A good arrangement that generates exactly one *good working relation*. This means that there are at least two boys sitting next to each other in the arrangement. These arrangements are called good arrangements of the first type. Let x_1 denote the total number of good arrangements of the first type.

(b) A good arrangement that generates exactly two *good working relations*. This means that boys and girls are sitting alternately. These arrangements are called good arrangements of the second type. Let x_2 denote the total number of good arrangements of the second type. Then $x_2 = 14! \times 15!$ as there are 14! ways to arrange all the boys around the desk and there are 15! ways to arrange all the girls each in a gap between two neighboring boys.

We have $x = x_1 + x_2$, where $x_2 = 14! \times 15!$ and

$$x_1 + 2x_2 = 14! \times 15! \times 2^{15}.$$

Therefore $x = 14! \cdot 15!(2^{15} - 1)$.

50. [Baltic Way 1999] Two squares on an 8×8 chessboard are called *touching* if they have at least one common vertex. Determine if it is possible for a king to begin in some square and visit all the squares exactly once in such a way that all moves except the first are made into squares touching an even number of squares already visited.

Solution: It is not possible for the king to visit all the squares. Assume for a contradiction that there exists a path such that all moves except the first are made into squares touching an even number of squares already visited. Clearly, the first move is made into a square touching exactly one square already visited, namely the starting square. Summing the number of touching squares previously visited over all the moves, we therefore obtain an odd number. On the other hand, every pair of touching squares is counted exactly once in this sum, by the member of the pair that was visited second. Thus, the sum is equal to the total number of touching pairs. But this number is even, since the numbers of touching pairs oriented north-south and east-west are equal, as are the numbers of touching pairs oriented northeast-southwest and northwest-southeast. Thus we have a contradiction, and no path exists.

51. [St. Petersburg 1988] A total of 119 residents live in a building with 120 apartments. We call an apartment *overpopulated* if there are at least 15 people living there. Every day the inhabitants of an overpopulated apartment have a quarrel and each goes off to a different apartment in the building (so they can avoid each other☺). Is it true that this process will necessarily be completed someday?

Solution: Let $p_1, p_2, \ldots, p_{120}$ denote the 120 apartments, and let a_i denote the number of residents in apartment p_i. We consider the quantity

$$S = \frac{a_1(a_1 - 1)}{2} + \frac{a_2(a_2 - 1)}{2} + \cdots + \frac{a_{120}(a_{120} - 1)}{2}.$$

(Assume that all the residents in an apartment shake hand with each other at the beginning of the day, then quantity S denotes the number of the handshakes in that day.) If all $a_i < 15$, then the process is completed and we are done. If not, without loss of generality, we assume that $a_1 \geq 15$ and that the inhabitants in p_1 go off to different apartments in the building. Assume that they go to apartments $p_{i_1}, p_{i_2}, \ldots, p_{i_{a_1}}$. On the next day, the quantity is changed by an amount of

$$a_{i_1} + a_{i_2} + \cdots + a_{i_{a_1}} - \frac{a_1(a_1 - 1)}{2},$$

which is positive as

$$a_{i_1} + a_{i_2} + \cdots + a_{i_{a_1}} \leq 119 - a_1 \leq 119 - 15 = 104$$

and

$$\frac{a_1(a_1 - 1)}{2} \geq \frac{15 \times 14}{2} = 105.$$

Hence the quantity is decreasing during this process. On the other hand, S starts as a certain finite number and S is nonnegative. Therefore this process has to be completed someday.

4

Solutions to Advanced Problems

1. [AIME 1985] In a tournament each player played exactly one game against each of the other players. In each game the winner was awarded 1 point, the loser got 0 points, and each of the two players earned 1/2 point if the game was a tie. After the completion of the tournament, it was found that exactly half of the points earned by each player were earned in games against the ten players with the least number of points. (In particular, each of the ten lowest scoring players earned half of her/his points against the other nine of the ten). What was the total number of players in the tournament?

   Solution: Assume that a total of n players participated in the tournament. We will obtain two expressions in n: one by considering the total number of points gathered by all of the players, and one by considering the number of points gathered by the losers (10 lowest scoring contestants) and those gathered by the winners (other $n - 10$ contestants) separately. To obtain the desired expressions, we will use that fact that if k players played against one another, then they played a total of $k(k-1)/2$ games, resulting in a total of $k(k-1)/2$ points to be shared among them. In view of the last observation, the n players gathered a total of $n(n-1)/2$ points in the tournament. Similarly, the losers had $10 \cdot 9/2$ or 45 points in games among themselves; since this accounts for half of their points, they must have had a total of 90 points. In games among themselves the $n - 10$ winners similarly gathered

$(n-10)(n-11)/2$ points; this also accounts for only half of their total number of points (the other half coming from games against the losers), so their total was $(n-10)(n-11)$ points. Thus we have the equation

$$n(n-1)/2 = 90 + (n-10)(n-11),$$

which is equivalent to

$$n^2 - 41n + 400 = 0.$$

Since the left member of this equation may be factored as $(n-16)(n-25)$, it follows that $n = 16$ or 25. We discard the first of these in view of the following observation: if there were only 16 players in the tournament, then there would have been only 6 winners, and the total of their points would have been 30 points, resulting in an average of 5 points for each of them. This is less than the 90/10 or 9 points gathered, on the average, by each of the losers! Therefore, $n = 25$; i.e., there were 25 players in the tournament.

Finally we show that such a tournament exists. Since $n = 25$, we have 15 winners and 10 losers. Every game that the winners play among themselves results in a tie, giving each winner $(15-1)/2 = 7$ points from games played with other winners. Likewise, all the games played among the losers result in ties, giving each of the 10 losers 4.5 points. For the ten games played by each winner against the losers, six are wins, two are losses, and two are ties, giving the winners another 7 points from games played with losers. This gives each loser three wins, nine losses, and three ties in games against winners, adding up to 4.5 more points. Thus each of the 25 players receives exactly half of his/her points in games against the losers, which is what we want.

2. [USAMO 1999 submission, Titu Andreescu] Let n be an odd integer greater than 1. Find the number of permutations p of the set $\{1, 2, \ldots, n\}$ for which

$$|p(1)-1| + |p(2)-2| + \cdots + |p(n)-n| = \frac{n^2-1}{2}.$$

Solution: We have

$$\begin{aligned} &|p(1)-1| + |p(2)-2| + \cdots + |p(n)-n| \\ = \quad &\pm 1 \pm 1 \pm 2 \pm 2 \pm \cdots \pm n \pm n. \end{aligned}$$

The maximum of $|p(1)-1|+|p(2)-2|+\cdots+|p(n)-n|$ is

$$2\left(-1-2-\cdots-\frac{n-1}{2}\right)-\frac{n+1}{2}+\frac{n+1}{2}$$

$$+2\left(\frac{n+3}{2}+\cdots+n\right)$$

$$= -\left(1+\frac{n-1}{2}\right)\frac{n-1}{2}+\left(\frac{n+3}{2}+n\right)\frac{n-1}{2}=\frac{n^2-1}{2}.$$

Let $p(\frac{n+1}{2})=k$. We must have

$$\left\{p(1),p(2),\ldots,p\left(\frac{n-1}{2}\right)\right\}=\left\{\frac{n+3}{2},\frac{n+5}{2},\ldots,n\right\}$$

and

$$\left\{p\left(\frac{n+3}{2}\right),p\left(\frac{n+5}{2}\right),\ldots,p(n)\right\}=\left\{1,2,\ldots,\frac{n+1}{2}\right\}-\{k\}$$

if $k\le\frac{n-1}{2}$, or

$$\left\{p(1),p(2),\ldots,p\left(\frac{n-1}{2}\right)\right\}=\left\{\frac{n+1}{2},\frac{n+3}{2},\ldots,n\right\}-\{k\}$$

and

$$\left\{p\left(\frac{n+3}{2}\right),p\left(\frac{n+5}{2}\right)\ldots,p(n)\right\}=\left\{1,2,\ldots,\frac{n-1}{2}\right\}$$

if $k\ge\frac{n+1}{2}$. There are

$$\frac{n-1}{2}\left[\left(\frac{n-1}{2}\right)!\right]^2+\frac{n+1}{2}\left[\left(\frac{n-1}{2}\right)!\right]^2=n\left[\left(\frac{n-1}{2}\right)!\right]^2$$

such permutations.

3. [AIME 1986] In a sequence of coin tosses one can keep a record of the number of instances when a tail is immediately followed by a head, a head is immediately followed by a head, etc. We denote these by TH, HH, etc. For example, in the sequence $HHTTHHHHTHHTTTT$ of 15 coin tosses we observe that there are five HH, three HT, two TH, and four TT subsequences. How many different sequences of 15 coin tosses will contain exactly two HH, three HT, four TH and five TT subsequences?

Solution: Think of such sequences of coin tosses as progressions of blocks of T's and H's, to be denoted by $\{T\}$ and $\{H\}$, respectively. Next note that each HT and TH subsequence signifies a transition from $\{H\}$ to $\{T\}$ and from $\{T\}$ to $\{H\}$, respectively. Since there should be three of the first kind and four of the second kind in each of the sequences of 15 coin tosses, one may conclude that each such sequence is of the form

$$\{T\}\{H\}\{T\}\{H\}\{T\}\{H\}\{T\}\{H\}. \tag{1}$$

Our next concern is the placement of $T's$ and $H's$ in their respective blocks, so as to assure that each sequence will have two HH and five TT subsequences. To this end, we will assume that each block in (1) initially contains only one member. Then, to satisfy the conditions of the problem, it will suffice to place 2 more H's into the $\{H\}$'s and 5 more T's into the $\{T\}$'s. Thus, to solve the problem, we must count the number of ways this can be accomplished.

Recall that the number of ways to put p indistinguishable balls (the extra H's and T's in our case) into q distinguishable boxes (the $\{H\}$'s and $\{T\}$'s, distinguished by their order in the sequence) is given by the formula $\binom{p+q-1}{p}$. (Students who are not familiar with this fact should verify it.) In our case, it implies that the 2 H's can be placed in the 4 $\{H\}$'s in $\binom{2+4-1}{2}$ or 10 ways, and the 5 T's can be placed in the 4 $\{T\}$'s in $\binom{5+4-1}{5}$ or 56 ways. The desired answer is the product, 560, of these numbers.

4. [IMO Shortlist 2001] Let $A = (a_1, a_2, \ldots, a_{2001})$ be a sequence of positive integers. Let m be the number of 3-element subsequences (a_i, a_j, a_k) with $1 \leq i < j < k \leq 2001$, such that $a_j = a_i + 1$ and $a_k = a_j + 1$. Considering all such sequences A, find the greatest value of m.

Solution: Consider the following two operations on the sequence A:

(1) If $a_i > a_{i+1}$, transpose these terms to obtain the new sequence $(a_1,\ a_2, \ldots, a_{i+1},\ a_i, \ldots, a_{2001})$.

(2) If $a_{i+1} = a_i + 1 + d$, where $d > 0$, increase $a_1, \ldots, a_i$ by d to obtain the new sequence
$(a_1{+}d,\ a_2{+}d, \ldots, a_i{+}d,\ a_{i+1}, \ldots, a_{2001})$.

It is clear that performing operation (1) cannot reduce m. By applying (1) repeatedly, the sequence can be rearranged to be nondecreasing. Thus we may assume that our sequence for which m is maximal is nondecreasing. Next, note that if A is nondecreasing, then performing operation (2) cannot reduce the value of m. It follows that any A with maximum m is of the form

$$(\underbrace{a,\ldots,a}_{t_1},\ \underbrace{a+1,\ldots,a+1}_{t_2},\ldots,\underbrace{a+s-1,\ldots,a+s-1}_{t_s})$$

where $t_1,\ldots,t_s$ are the number of terms in each subsequence, and $s \geq 3$. For such a sequence A,

$$m = t_1t_2t_3 + t_2t_3t_4 + \cdots + t_{s-2}t_{s-1}t_s. \qquad (*)$$

It remains to find the best choice of s and the best partition of 2001 into positive integers $t_1,\ldots,t_s$.

The maximum value of m occurs when $s = 3$ or $s = 4$. If $s > 4$ then we may increase the value given by $(*)$ by using a partition of 2001 into $s - 1$ parts, namely

$$t_2, t_3, (t_1 + t_4), \ldots, t_s.$$

Note that when $s = 4$ this modification does not change the value given by $(*)$. Hence the maximum value m can be obtained with $s = 3$. In this case, $m = t_1t_2t_3$ is largest when $t_1 = t_2 = t_3 = 2001/3 = 667$. Thus the maximum value of m is 667^3. This maximum value is attained when $s = 4$ as well, in this case for sequences with $t_1 = a$, $t_2 = t_3 = 667$, and $t_4 = 667 - a$, where $1 \leq a \leq 666$.

5. [USAMO 1989 submission, Paul Zeitz] Twenty-three people of positive integral weights decide to play football. They select one person as referee and then split up into two 11-person teams of equal total weights. It turns out that no matter who the referee this can always be done. Prove that all 23 people have equal weights.

Solution: Assume on the contrary that there is a set of 23 not all equal integer weights satisfying the conditions of the problem. Then among such sets there is a set $A = (a_1, a_2, \ldots, a_{23})$ with the smallest total weight $w = a_1 + a_2 + \cdots + a_{23}$. If a_i is the referee, then $w - a_i = 2s_i$, where s_i is the total weight of each team. Hence $a_i \equiv w \pmod 2$, that is, a_i's have the same parity.

If the a_i's are all even, we can replace A by

$$A' = \left(\frac{a_1}{2}, \frac{a_2}{2}, \ldots, \frac{a_{23}}{2}\right)$$

a set of less total weight that satisfies the conditions of the problem. And since the a_i's are not all equal, the $a_i/2$'s are not all equal. This contradicts the fact that A is such a set with minimum total weight.

If the a_i's are all odd, we can use

$$A'' = ((a_1 + 1)/2, (a_2 + 1)/2, \ldots, (a_{23} + 1)/2)$$

to lead to a similar contradiction.

Hence our assumption was wrong and all 23 people must have equal weights.

Note: With a bit more knowledge on matrix theory, one can show a more general work. Let n be a positive integer, and let $x_1, x_2, \ldots, x_{2n+1}$ be real numbers. If any one of them is removed, the remaining ones can be divided into two sets of n numbers with equal sums, then

$$x_1 = x_2 = \cdots + x_{2n+1}.$$

6. [IMO Shortlist 1998] Determine the smallest integer n, $n \geq 4$, for which one can choose four different numbers a, b, c, d from any n distinct integers such that $a + b - c - d$ is divisible by 20.

Solution: We first consider only sets of integers with distinct residues modulo 20. For such a set of k elements, there are a total of $k(k-1)/2$ pairs. Therefore, if $k(k-1)/2 > 20$ (i.e., $k \geq 7$), then there exist two pairs of numbers (a, b) and (c, d) such that $a + b \equiv c + d \pmod{20}$ and a, b, c, d are all distinct.

In general, let us consider a set of 9 distinct integers. If there are seven of them that have distinct residues modulo 20, we are done by the above argument. Suppose that there are at most 6 distinct residues modulo 20 in this set, i.e., at least 3 residues have to be repeated. Then either there are 4 numbers (a, b, c, d) with $a \equiv b \equiv c \equiv d \pmod{20}$ or there are 2 pairs of numbers (a, c) and (b, d) with $a \equiv c$ and $b \equiv d \pmod{20}$. In either case, we can find a desired quadruple (a, b, c, d).

It is not difficult to find a set of 8 numbers which does not have the property we want:

$$\{0, 20, 40, 1, 2, 4, 7, 12\}.$$

Residues of these numbers modulo 20 are 0, 0, 0, 1, 2, 4, 7, 12, respectively. These residues have the property that each nonzero residue is greater than the sum of any two smaller ones, and the sum of any two is less than 20. Let a, b, c, d be the respective residues of 4 distinct numbers of this set. Without loss of generality, we may assume that a is the largest (as a is interchangeable with b and can interchange with either c or d by multiplication by -1, which does not affect its divisibility by 20). Thus a is a nonzero residue and

$$0 < a - c - d \leq a + b - c - d \leq a + b < 20.$$

Hence $a + b - c - d$ is not divisible by 20.

Therefore the desired minimum value of n is 9.

7. [AIME 2001] A mail carrier delivers mail to the nineteen houses on the east side of Elm Street. The carrier notes that no two adjacent houses ever get mail on the same day, but that there are never more than two houses in a row that get no mail on the same day. How many different patterns of mail delivery are possible?

First Solution: The first condition implies that at most ten houses get mail in one day, while the second condition implies that at least six houses get mail. If six houses get mail, they must be separated from each other by a total of at least five houses that do not get mail. The other eight houses that do not get mail must be distributed in the seven spaces on the sides of the six houses that do get mail. This can be done in 7 ways: put two at each end of the street and distribute the other four in $\binom{5}{4} = 5$ ways, or put one in each of the seven spaces and an extra one at one end of the street or the other. If seven houses get mail, they create eight spaces, six of which must contain at least one house that does not get mail. The remaining six houses that do not get mail can be distributed among these eight spaces in 113 ways: six of the eight spaces can be selected to receive a single house in $\binom{8}{6} = 28$ ways; two houses can be placed at each end of the street and two intermediate spaces be selected in $\binom{6}{2} = 15$ ways; and two houses can be placed at one end of the street and four spaces selected for a single house in $2\binom{7}{4} = 70$ ways. Similar reasoning shows that there are $\binom{9}{4} + 1 + 2\binom{8}{2} = 183$ patterns when eight houses get mail, and $2 + \binom{10}{2} = 47$ patterns when nine houses get mail. When ten houses get mail, there is only one pattern, and thus the total number of patterns is $7 + 113 + 183 + 47 + 1 = 351$.

Second Solution: Consider n-digit strings of zeros and ones, which represent no mail and mail, respectively. Such a sequence is called *acceptable* if it contains no occurrences of 11 or 000. Let f_n be the number of acceptable n-digit strings, let a_n be the number of acceptable n-digit strings in which 00 follows the leftmost 1, and let b_n be the number of acceptable n-digit strings in which 01 follows the leftmost 1. Notice that $f_n = a_n + b_n$ for $n \geq 5$. Deleting the leftmost occurrence of 100 shows that $a_n = f_{n-3}$, and deleting 10 from the leftmost occurrence of 101 shows that $b_n = f_{n-2}$. It follows that $f_n = f_{n-2} + f_{n-3}$ for $n \geq 5$. It is straightforward to verify the values $f_1 = 2$, $f_2 = 3$, $f_3 = 4$, and $f_4 = 7$. Then the recursion can be used to find that $f_{19} = 351$.

8. [China 1996] For $i = 1, 2, \ldots, 11$, let M_i be a set of five elements, and assume that for every $1 \le i < j \le 11$, $M_i \cap M_j \ne \emptyset$. Let m be the largest number for which there exist $M_{i_1}, \ldots, M_{i_m}$ among the chosen sets with $\cap_{k=1}^{m} M_{i_k} \ne \emptyset$. Find the minimum value of m over all possible initial choices of M_i.

   Solution: The minimum value of m is 4.

   We first show that $m \ge 4$. Let $X = \cup_{i=1}^{11} M_i$, and for each $x \in X$, let $n(x)$ denote the number of i's such that $x \in M_i$, $1 \le i \le 11$. Then $m = \max\{n(x), x \in X\}$. Note that

$$\sum_{x \in X} n(x) = 55.$$

   Since $M_i \cap M_j \ne \emptyset$, there are $\binom{11}{2} = 55$ nonempty intersections. On the other hand, each element x appears in $\binom{n(x)}{2}$ intersections. Therefore,

$$\sum_{x \in X} \binom{n(x)}{2} \ge \binom{11}{2} = 55.$$

   It follows that

$$\sum_{x \in X} \frac{n(x)(n(x) - 1)}{2} \ge 55,$$

   implying that

$$\frac{m-1}{2} \sum_{x \in X} n(x) \ge 55.$$

   Hence $\frac{m-1}{2} \ge 1$, or $m \ge 3$. If $m = 3$, then all equalities hold; more precisely, $n(x) = m = 3$ for all x. But since $\sum_{x \in X} n(x) = 55$ and 55 is not divisible by 3, $n(x)$ cannot always equal 3. Therefore $m \ge 4$.

   Now we prove that $m = 4$ can be obtained. We consider the following 4×4 array:

$$\begin{array}{cccc} a & b & c & d \\ e & f & g & h \\ 1 & 2 & 3 & 4 \\ 5 & 6 & 7 & 8 \end{array}$$

   It is not difficult to see that the sets $M_1 = \{a, b, c, d, H\}$, $M_2 = \{e, f, g, h, H\}$, $M_3 = \{1, 2, 3, 4, H\}$, $M_4 = \{5, 6, 7, 8, H\}$ (we call them *horizontal* sets); $M_5 = \{a, e, 1, 5, V\}$, $M_6 = \{b, f, 2, 6, V\}$, $M_7 = \{c, g, 3, 7, V\}$, $M_8 = \{d, h, 4, 8, V\}$ (we call them *vertical* sets); $M_9 = \{a, f, 3, 8, D\}$, $M_{10} = \{b, g, 4, 5, D\}$, $M_{11} = \{c, h, 1, 6, D\}$ (we call them *diagonal* sets); satisfy the conditions of the problem with $m = 4$.

9. [AIME 1998] Define a *domino* to be an ordered pair of *distinct* positive integers. A *proper sequence* of dominos is a list of distinct dominos in which the first coordinate of each pair after the first equals the second coordinate of the immediately preceding pair, and in which (i, j) and (j, i) do not *both* appear for any i and j. Let D_{40} be the set of all dominos whose coordinates are no larger than 40. Find the length of the longest proper sequence of dominos that can be formed using the dominos of D_{40}.

First Solution: Let $A_n = \{1, 2, 3, \ldots, n\}$ and D_n the set of dominos that can be formed using integers in A_n. Each k in A_n appears in $2(n-1)$ dominos in D_n; hence it appears at most $n-1$ times in a proper sequence from D_n. Except possibly for the integers i and j that begin and end a proper sequence, every integer appears an even number of times in the sequence. Thus, if n is even, every integer other than i and j appears in at most $n-2$ dominos. This gives an upper bound of

$$\frac{1}{2}[(n-2)^2 + 2(n-1)] = \frac{n^2 - 2n + 2}{2}$$

dominos in the longest proper sequence in D_n. This bound is in fact attained for every even n. It is easy to verify this for $n = 2$, so assume inductively that a sequence of this length has been found for a particular value of n. Without loss of generality, assume $i = 1$ and $j = 2$, and let ${}_pX_{p+2}$ denote a four-domino sequence of the form $(p, n+1)(n+1, p+1)(p+1, n+2)(n+2, p+2)$. By appending

$${}_2X_4\,,\ {}_4X_6\,,\ \ldots\,,\ {}_{n-2}X_n\,,\ (n, n+1)(n+1, 1)(1, n+2)(n+2, 2)$$

to the given proper sequence, a proper sequence of length

$$\begin{aligned} \frac{n^2 - 2n + 2}{2} + 4 \cdot \frac{n-2}{2} + 4 &= \frac{n^2 + 2n + 2}{2} \\ &= \frac{(n+2)^2 - 2(n+2) + 2}{2} \end{aligned}$$

is obtained that starts at $i = 1$ and ends at $j = 2$. This completes the inductive proof. In particular, the longest proper sequence when $n = 40$ is 761.

Second Solution: A proper sequence can be represented by writing the common coordinates of adjacent ordered pairs once. For example, represent $(4, 7), (7, 3), (3, 5)$ as $4, 7, 3, 5$. Label the vertices of a regular n-gon $1, 2, 3, \ldots, n$. Each domino is thereby represented by a directed segment from one vertex of the n-gon to another, and a proper sequence is represented as a path that retraces no segment. Each time such a path reaches a non-terminal vertex, it must leave it. Thus, when n is even, it is not possible for such a path to trace every segment, since an odd number of segments emanate from each vertex. By removing $\frac{1}{2}(n-2)$ suitable segments, however, it can be arranged that $n-2$ segments will emanate from $n-2$ of the vertices, and that an odd number of segments will emanate from exactly two of the vertices. In this situation, a path can be found that traces every remaining segment exactly once, starting at one of the two exceptional vertices and finishing at the other. This path will have length $\binom{n}{2} - \frac{1}{2}(n-2)$, which is 761 when $n = 40$.

Note: When n is odd, a proper sequence of length $\binom{n}{2}$ can be found using the dominos of D_n. In this case, the second coordinate of the final domino equals the first coordinate of the first domino. In the language of graph theory, this is an example of an *Eulerian circuit.*

10. [High School Mathematics, 1994/1, Qihong Xie] Find the number of subsets of $\{1, \ldots, 2000\}$, the sum of whose elements is divisible by 5.

Solution: The answer is $\frac{1}{5}(2^{2000} + 2^{402})$.

Consider the polynomial

$$f(x) = (1+x)(1+x^2)\cdots(1+x^{2000}).$$

Then there is a bijection between each subset $\{a_1, a_2, \ldots, a_m\}$ of $\{1, 2, \ldots, 2000\}$ and the term $x^{a_1}x^{a_2}\cdots x^{a_m}$. Hence we are looking for the sum of coefficients of terms x^{5k} in $f(x)$, k a positive integer. Let S denote that sum.

Let $\xi = e^{2\pi i/5}$ be a 5þ root of unity. Then $\xi^5 = 1$ and $1+\xi+\xi^2+\xi^3+\xi^4 = 0$. Hence

$$S = \frac{1}{5}\sum_{j=1}^{5} f(\xi^j).$$

Note that $\xi, \xi^2, \xi^3, \xi^4, \xi^5 = 1$ are the roots of $g(x) = x^5 - 1$, that is

$$g(x) = x^5 - 1 = (x-\xi)(x-\xi^2)(x-\xi^3)(x-\xi^4)(x-\xi^5).$$

It follows that

$$g(-1) = -2 = (-1-\xi)(-1-\xi^2)(-1-\xi^3)(-1-\xi^4)(-1-\xi^5).$$

Therefore

$$(1+\xi)(1+\xi^2)(1+\xi^3)(1+\xi^4)(1+\xi^5) = 2$$

and $f(\xi) = 2^{400}$. Likewise, $f(\xi^j) = 2^{400}$ for $j = 2, 3, 4$. Finally, we calculate $f(\xi^5) = f(1) = 2^{2000}$. We obtain

$$S = \frac{1}{5}\left(4 \cdot 2^{400} + 2^{2000}\right) = \frac{1}{5}\left(2^{402} + 2^{2000}\right).$$

11. [MOSP 1999] Let X be a finite set of positive integers and A a subset of X. Prove that there exists a subset B of X such that A equals the set of elements of X which divide an odd number of elements of B.

Solution: We construct B in stages. Set $B = \emptyset$ and consider every number in X, starting with the largest and going down. For each element $x \in X$, see whether it divides the correct parity of elements in B. (That is, if $x \in A$, x divides an odd number of elements in B; if $x \in X - A$, x divides an even number of elements in B.) If it does not, add it to B. Thus the first element added to B is the largest element of A. Now, this procedure will not change the divisibility condition for any element greater than x, and will fulfill the condition for x. Thus when all elements of X have been examined, the divisibility conditions will be satisfied by all elements of X, and B will be as desired.

12. [AIME 2000] A stack of 2000 cards is labeled with the integers from 1 to 2000, with different integers on different cards. The cards in the stack are not in numerical order. The top card is removed from the stack and placed on the table, and the next card in the stack is moved to the bottom of the stack. The new top card is removed from the stack and placed on the table, to the right of the card already there, and the next card in the stack is moved to the bottom of the stack. This process—placing the top card to the right of the cards already on the table and moving the next card in the stack to the bottom of the stack—is repeated until all cards are on the table. It is found that, reading left to right, the labels on the cards are now in ascending order: $1, 2, 3, \ldots, 1999, 2000$. In the original stack of cards, how many cards were above the card labeled 1999?

First Solution: Run the process backwards. Start by picking up the card labeled 2000. Next, pick up the card labeled 1999, place it on top of the stack, and bring the bottom card to the top of the stack. Next pick up the card labeled 1998, place it on top of the stack, and bring the bottom card to the top of the stack. The card labeled 1999 is now at the top of a three-card stack. Note that the top card of an m-card stack will become the top card of a $2m$-card stack after m more cards have been picked up (and m cards have been moved from the bottom of the stack to the top). It follows by induction that the card labeled 1999 is the top card when the number of cards in the stack is $3 \cdot 2^k$ for any nonnegative integer k that satisfies $3 \cdot 2^k < 2000$. In particular, the last time that this happens is just after $3 \cdot 2^9 = 1536$ cards have been picked up. The cards remaining on the table are labeled 1 through 464. After each of the cards labeled $464, 463, \ldots, 2$ is picked up and placed on top of the stack, another card is brought from the bottom of the stack to the top. Finally, the card labeled 1 is placed on top of the stack and the stack is in its original state. This puts $2 \cdot 463 + 1 = 927$ cards on top of the card labeled 1999.

Solution: Because the process causes the cards on the table to appear in ascending order, the card labeled 1999 is next-to-last placed on the table. To keep track of that card, first notice that, when a stack of 2^m cards is dealt in this way, the next-to-last card placed on the table begins at position 2^{m-1} in the stack; then apply the process to a stack of $2^{11} = 2048$ cards. After 48 of the cards have been placed on the table and 48 more cards have been moved from the top of the stack to the bottom, a 2000-card stack remains. Remove the cards that are on the table. The next-to-last card that will be placed on the table from the 2000-card stack is the card that began at position 1024 in the 2048-card stack. The position of that card in the 2000-card stack is $1024 - (48 + 48) = 928$, so the number of cards above it is 927.

13. Form a 2000×2002 screen with unit screens. Initially, there are more than 1999×2001 unit screens which are *on*. In any 2×2 screen, as soon as there are 3 unit screens which are *off*, the 4þ screen turns off automatically. Prove that the whole screen can never be totally off.

    Solution: For a screen to turn off, it has to be the 4þ screen of a 2×2 screen with the other 3 screens off. Conversely, each 2×2 subscreen can be used only once to turn off a screen. Since there are 1999×2001 2×2 subscreens, at most 1999×2001 screens can be turned off. Hence the whole screen can never be totally off.

14. [AIME 1988] In an office, at various times during the day, the boss gives the secretary a letter to type, each time putting the letter on top of the pile in the secretary's inbox. When there is time, the secretary takes the top letter off the pile and types it. There are nine letters to be typed during the day, and the boss delivers them in the order 1, 2, 3, 4, 5, 6, 7, 8, 9. While leaving for lunch, the secretary tells a colleague that letter 8 has already been typed, but says nothing else about the morning's typing. The colleague wonders which of the nine letters remain to be typed after lunch and in what order they will be typed. Based upon the above information, how many such *after lunch typing orders* are possible? (That there are no letters left to be typed is one of the possibilities.)

Solution: At any given time, the letters in the box are in decreasing order from top to bottom. Thus the sequence of letters in the box is uniquely determined by the set of letters in the box. We have two cases: letter 9 arrived before lunch or it did not.

- *Case 1:* Since letter 9 arrived before lunch, no further letters will arrive, and the number of possible orders is simply the number of subsets of $T = 1, 2, \ldots, 6, 7, 9$ which might still be in the box. In fact, each subset of T is possible, because the secretary might have typed letters not in the subset as soon as they arrived and not typed any others. Since T has 8 elements, it has $2^8 = 256$ subsets (including the empty set).
- *Case 2:* Since letter 9 didn't arrive before lunch, the question is: where can it be inserted in the typing order? Any position is possible for each subset of $U = \{1, 2, \ldots, 6, 7\}$ which might have been left in the box during lunch (in descending order). For instance, if the letters in the box during lunch are 6, 3, 2 then the typing order 6, 3, 9, 2 would occur if the boss would deliver letter 9 just after letter 3 was typed. There would seem to be $k + 1$ places at which letter 9 could be inserted into a sequence of k letters. However, if letter 9 is inserted at the beginning of the sequence (i.e., at the top of the pile, so it arrives before any after lunch typing is done), then we are duplicating an ordering from Case 1. Thus, if k letters are in the basket after returning from lunch, there are k places to insert letter 9 (without duplicating Case 1 orderings). Thus we obtain
$$\sum_{k=0}^{7} k\binom{7}{k} = 7(2^{7-1}) = 448$$
new orderings in Case 2.

Combining these cases gives $256 + 448 = 704$ possible typing orders.

15. [China 1994, Wushang Shu] Let n be a positive integer. Prove that
$$\sum_{k=0}^{n} 2^k \binom{n}{k}\binom{n-k}{\lfloor (n-k)/2 \rfloor} = \binom{2n+1}{n}.$$

First Solution: (By Chenchang Zhu) For a polynomial $p(x)$, let $[x^n](p(x))$ be the coefficient of the term x^n in $p(x)$. Consider the polynomial $p(x) = (x+1)^{2n}$. It is easy to see that
$$[x^{n-1}](p(x)) + [x^n](p(x)) = \binom{2n}{n-1} + \binom{2n}{n} = \binom{2n+1}{n}.$$

Hence it suffices to show that

$$[x^{n-1}](p(x)) + [x^n](p(x)) = \sum_{k=0}^{n} 2^k \binom{n}{k} \binom{n-k}{\lfloor (n-k)/2 \rfloor}.$$

Note that

$$\begin{aligned} p(x) &= (x+1)^{2n} = (x^2+2x+1)^n = \sum_{i+j+k=n} \frac{n!}{i!j!k!}(x^2)^i(2x)^j \\ &= \sum_{0 \le i+j \le n} \frac{n!}{i!j!(n-i-j)!} 2^j x^{2i+j}, \end{aligned}$$

where i, j, k are nonnegative integers. Hence

$$\begin{aligned} &[x^{n-1}](p(x)) + [x^n](p(x)) \\ &= \sum_{\substack{0 \le i+j \le n \\ 2i+j=n}} \frac{n!}{i!j!(n-i-j)!} \cdot 2^j + \sum_{\substack{0 \le i+j \le n \\ 2i+j=n-1}} \frac{n!}{i!j!(n-i-j)!} \cdot 2^j \\ &= \sum_{\substack{0 \le i+(n-2i) \le n \\ 0 \le i, n-2i}} \frac{n!}{i!(n-2i)!i!} \cdot 2^{n-2i} \\ &\quad + \sum_{\substack{0 \le i+(n-2i-1) \le n \\ 0 \le i, n-2i-1}} \frac{n!}{i!(n-2i-1)!(i+1)!} \cdot 2^{n-2i-1} \\ &= \sum_{i=1}^{\lfloor \frac{n}{2} \rfloor} \binom{n}{2i}\binom{2i}{i} 2^{n-2i} + \sum_{i=1}^{\lfloor \frac{n-1}{2} \rfloor} \binom{n}{2i+1}\binom{2i+1}{i} 2^{n-2i-1} \\ &= \sum_{\substack{s=0 \\ s \text{ even}}}^{n} \binom{n}{s}\binom{s}{\lfloor \frac{s}{2} \rfloor} 2^{n-s} + \sum_{\substack{s=1 \\ s \text{ odd}}}^{n} \binom{n}{s}\binom{s}{\lfloor \frac{s}{2} \rfloor} 2^{n-s} \\ &= \sum_{s=0}^{n} \binom{n}{s}\binom{s}{\lfloor \frac{s}{2} \rfloor} 2^{n-s} \end{aligned}$$

Setting $n - s = k$ in the last expression gives

$$\begin{aligned}[x^{n-1}](p(x)) + [x^n](p(x)) &= \sum_{k=0}^{n} \binom{n}{n-k} \binom{n-k}{\lfloor \frac{n-k}{2} \rfloor} 2^k \\ &= \sum_{k=0}^{n} \binom{n}{k} \binom{n-k}{\lfloor \frac{n-k}{2} \rfloor} 2^k,\end{aligned}$$

as desired.

Second Solution: (By Jian Gu) We consider a combinatorial model. There are $2n$ students, n boys and n girls, in a class with their teacher T. Let $g_1, g_2, \ldots, g_n$ denote all the girls, and let $b_1, b_2, \ldots, b_n$ denote all the boys. For $1 \le i \le n$, students (g_i, b_i) are paired. The class has n tickets to an exciting soccer game.

We consider the number of ways to find n people to go to the game. The obvious answer is

$$\binom{2n+1}{n}.$$

On the other hand, we also can calculate this number in the following way. For any fixed integer k, $1 \le k \le n$, we find k pairs from the n pairs of students and give each pair 1 ticket. There are $\binom{n}{k}2^k$ ways to find k pairs and pick one student from each pair to go to the game. We have $n-k$ tickets left and $n-k$ pairs of student left. We pick $\lfloor \frac{n-k}{2} \rfloor$ pairs and give each of those pairs 2 tickets. There are

$$\binom{n-k}{\lfloor \frac{n-k}{2} \rfloor}$$

ways to do so. Now we have already assigned $S = k + 2\lfloor \frac{n-k}{2} \rfloor$ tickets. If $n-k$ is odd, $S = n-1$ and we assign the last ticket to the teacher T; if $n-k$ is even, $S = n$ and we have assigned all the tickets already. It is not difficult to see that as k takes all the values from 1 to n, we obtain all possible ways of assigning the n tickets. Therefore, there are

$$\sum_{k=0}^{n} 2^k \binom{n}{k} \binom{n-k}{\lfloor (n-k)/2 \rfloor}$$

ways to find n people to go the game. Hence

$$\sum_{k=0}^{n} 2^k \binom{n}{k} \binom{n-k}{\lfloor (n-k)/2 \rfloor} = \binom{2n+1}{n},$$

as desired.

16. [Bay Area Math Circle 1999] Let m and n be positive integers. Suppose that a given rectangle can be tiled by a combination of horizontal $1 \times m$ strips and vertical $n \times 1$ strips. Prove that it can be tiled using only one of the two types.

    Solution: Assume that the dimensions of the rectangle are $a \times b$. It is clear that both a and b are positive integers. We want to show that either a is divisible by m or b is divisible by n. Let $\zeta = \operatorname{cis} 2\pi/m$ and $\xi = \operatorname{cis} 2\pi/n$ be the mþ and nþ roots of unity, respectively. Divide the rectangle into ab unit squares, and write the number $\zeta^x \xi^y$ in the square in the xþ column and yþ row. For each vertical strip, the sum of the numbers written in it is

$$\zeta^x \xi^y (1 + \xi + \xi^2 + \cdots + \xi^{n-1}) = \zeta^x \xi^y \cdot \frac{\xi^n - 1}{\xi - 1} = 0.$$

    Likewise the sum of the numbers in any horizontal strip is also 0. Since the rectangle is tiled by these strips, the sum of all the numbers in the rectangle is 0. But this sum is equal to

$$(\zeta + \zeta^2 + \cdots + \zeta^a)(\xi + \xi^2 + \cdots + \xi^b) = \zeta\xi \cdot \frac{\zeta^a - 1}{\zeta - 1} \cdot \frac{\xi^b - 1}{\xi - 1}.$$

    Therefore we must have $\zeta^a = 1$ or $\xi^b = 1$, implying that $m \mid a$ or $n \mid b$, respectively.

17. Given an initial sequence $a_1, a_2, \ldots, a_n$ of real numbers, we perform a series of steps. At each step, we replace the current sequence $x_1, x_2, \ldots, x_n$ with $|x_1 - a|, |x_2 - a|, \ldots, |x_n - a|$ for some a. For each step, the value of a can be different.

    (a) Prove that it is always possible to obtain the null sequence consisting of all 0's.

    (b) Determine with proof the minimum number of steps required, regardless of initial sequence, to obtain the null sequence.

    Solution: First we show that n steps are enough to obtain the null sequence. Let $(a_1^{(k)}, a_2^{(k)}, \ldots, a_n^{(k)})$ denote the sequence after the kþ step, and let $a^{(k)}$ be the value of the number a chosen to be subtracted in the kþ step. We set $a^{(1)} = \frac{a_1+a_2}{2}$. Hence

$$a_1^{(1)} = a_2^{(1)} = \frac{1}{2}|a_1 - a_2|.$$

We then take $a^{(2)} = \frac{a_1^{(1)}+a_3^{(1)}}{2}$ to obtain

$$a_1^{(2)} = a_2^{(2)} = a_3^{(2)} = \frac{|a_1^{(1)} - a_3^{(1)}|}{2},$$

and so on. At the kþ step, we take $a^{(k)} = \frac{a_1^{(k-1)}+a_{k+1}^{(k-1)})}{2}$ to obtain

$$a_1^{(k)} = a_2^{(k)} = \cdots = a_{k+1}^{(k)} = \frac{|a_1^{(k-1)} - a_{k+1}^{(k-1)}|}{2}.$$

In this way, we obtain a sequence $a_1^{(n-1)}, a_2^{(n-1)}, \ldots, a_n^{(n-1)}$ with $a_1^{(n-1)} = a_2^{(n-1)} = \cdots = a_n^{(n-1)}$ after $n-1$ steps. At the nþ step we take $a^{(n)} = a_1^{(n-1)}$ to obtain the null sequence.

We prove that n steps are necessary for the initial sequence $1, 2!, 3!, \ldots, n!$ by induction on n. The base case $n = 1$ is trivial.

Assume that our statement is true for some positive integer k, that is, we need at least k steps to turn the sequence $1, 2!, 3!, \ldots, k!$ into the null sequence. We show that we need at least $k+1$ steps to turn the sequence $1, 2!, \ldots, (k+1)!$ into the null sequence.

The key observation is that if m is the minimum number of steps required to turn the sequence $a_1, a_2, \ldots, a_n$ into the null sequence, then

$$m^{(k-1)} \le a^{(k)} \le M^{(k-1)}$$

for $1 \le k \le m$, where

$$m^{(k-1)} = \min\{a_1^{(k-1)}, a_2^{(k-1)}, \ldots, a_n^{(k-1)}\}$$

and

$$M^{(k-1)} = \max\{a_1^{(k-1)}, a_2^{(k-1)}, \ldots, a_n^{(k-1)}\}.$$

Indeed, if $a^{(k)} < m^{(k-1)}$, then for all i,

$$\begin{aligned} a_i^{(k+1)} &= |a_i^{(k)} - a^{(k+1)}| \\ &= ||a_i^{(k-1)} - a^{(k)}| - a^{(k+1)}| \\ &= |a_i^{(k-1)} - a^{(k)} - a^{(k+1)}| \\ &= |a_i^{(k-1)} - (a^{(k)} + a^{(k+1)})|. \end{aligned}$$

We can set the value of a at the kþ step to $a^{(k)} + a^{(k+1)}$ to save a step, which contradicts the fact that m is the minimum number of steps needed. On the other hand, if $a^{(k)} > M^{(k-1)}$, then for all i,

$$\begin{aligned} a_i^{(k+1)} &= |a_i^{(k)} - a^{(k+1)}| \\ &= ||a_i^{(k-1)} - a^{(k)}| - a^{(k+1)}| \\ &= |-a_i^{(k-1)} + a^{(k)} - a^{(k+1)}| \\ &= |a_i^{(k-1)} - (a^{(k)} - a^{(k+1)})|. \end{aligned}$$

We can set the value of a at the kþ step to $a^{(k)} - a^{(k+1)}$ to save a step, which contradicts the fact that m is the minimum number of steps needed.

It follows from the above argument that $M^{(0)} \geq M^{(1)} \geq \cdots \geq M^{(m)}$, and hence $a^{(k)} \leq M^{(0)}$ for all k. Note also that since $m^{(k)}$ is always nonnegative, $a^{(k)} \geq 0$ for all k.

We are now ready to prove our inductive step. We approach indirectly by assuming that it is possible to turn the sequence $1, 2!, \ldots, (k+1)!$ into the null sequence in k steps. Then the subsequence $1, 2!, \ldots, k!$ has also been turned into the null sequence. By the induction hypothesis, k is the minimum number of steps needed to turn the sequence $1, 2!, \ldots, k!$ into the null sequence. By our observation above, we conclude that $0 \leq a^{(i)} \leq k!$ for $1 \leq i \leq k$. But then

$$\begin{aligned} a_{k+1}^{(k)} &= ||\cdots|(k+1)! - a^{(1)}| - a^{(2)}| - \cdots - a^{(k)}| \\ &= (k+1)! - (a^{(1)} + a^{(2)} + \cdots + a^{(k)}) \\ &\geq (k+1)! - k \cdot k! > 0, \end{aligned}$$

which contradicts the fact that $a_{k+1}^{(k)} = 0$ as it is part of the null sequence. Therefore our assumption was wrong and we need at least $k+1$ steps to turn the sequence $1, 2!, \ldots, (k+1)!$ into the null sequence. Our induction is thus complete.

18. [China 2000, Yuming Huang] The sequence $\{a_n\}_{n\geq 1}$ satisfies the conditions $a_1 = 0, a_2 = 1$,

$$a_n = \frac{1}{2}na_{n-1} + \frac{1}{2}n(n-1)a_{n-2} + (-1)^n\left(1 - \frac{n}{2}\right),$$

$n \geq 3$. Determine the explicit form of

$$\begin{aligned} f_n &= a_n + 2\binom{n}{1}a_{n-1} + 3\binom{n}{2}a_{n-2} \\ &\quad + \cdots + (n-1)\binom{n}{n-2}a_2 + n\binom{n}{n-1}a_1. \end{aligned}$$

First Solution: It is straightforward to show by induction that

$$a_n = na_{n-1} + (-1)^n,$$

which implies that

$$a_n = n! - \frac{n!}{1!} + \frac{n!}{2!} - \frac{n!}{3!} + \cdots + (-1)^n \frac{n!}{n!},$$

or

$$a_n = n!\left(1 - \frac{1}{1!} + \frac{1}{2!} - \frac{1}{3!} + \cdots + (-1)^n \frac{1}{n!}\right).$$

Therefore, by Bernoulli–Euler's famous formula of misaddressed letters, a_n is the number of derangements of $(1, 2, \ldots, n)$, i.e., the number of permutations of this n-tuple with no fixed points.

Then f_n can be interpreted as follows: For each non-identity permutation of $(1, 2, \ldots, n)$, gain one mark; then gain one mark for each fixed point of the permutation. Then f_n is the total mark scored by all the non-identity permutations.

On the other hand, the total mark can also be calculated as a sum of the marks of each element gained in all the non0identity permutations. There are $n! - 1$ non-identity permutations and each number is fixed in $(n-1)! - 1$ non-identity permutations for a total of

$$f_n = n! - 1 + n((n-1)! - 1) = 2 \cdot n! - n - 1$$

marks.

Second Solution: We present another method proving that a_n is the number of derangements of $(1, 2, \ldots, n)$. We have

$$\begin{aligned} a_n &= na_{n-1} + (-1)^n = a_{n-1} + (n-1)a_{n-1} + (-1)^n \\ &= [(n-1)a_{n-2} + (-1)^{n-1}] + (n-1)a_{n-1} + (-1)^n \\ &= (n-1)(a_{n-1} + a_{n-2}). \end{aligned}$$

Now let b_n be the number of derangements of $(1, 2, \ldots, n)$. In a derangement, either

(a) 1 maps to k and k maps to 1 for some $k \neq 1$. Then there are $(n-1)$ possible values for k and for each k there are b_{n-2} derangements of the other $n-2$ elements. Hence there are $(n-1)b_{n-2}$ such derangements.

(b) 1 maps to k and k maps to m for some $k, m \neq 1$. Note that $k \neq m$. Then this is simply a derangement of $(2, \ldots, k-1, 1, k+1, \ldots n)$ with 1 mapping to m. Again there are $n-1$ possible values for k and for each k there are b_{n-1} derangements. Hence there are $(n-1)b_{n-1}$ such derangements.

Therefore $b_n = (n-1)(b_{n-1} + b_{n-2})$. Since $a_1 = b_1 = 0$ and $a_2 = b_2 = 1$, $a_n = b_n$, as claimed.

19. [USAMO 2000 submission, Richard Stong] For a set A, let $|A|$ and $s(A)$ denote the number of the elements in A and the sum of elements in A, respectively. (If $A = \emptyset$, then $|A| = s(A) = 0$.) Let S be a set of positive integers such that

(a) there are two numbers $x, y \in S$ with $\gcd(x, y) = 1$;

(b) for any two numbers $x, y \in S$, $x + y \in S$.

Let T be the set of all positive integers not in S. Prove that $s(T) \leq |T|^2 < \infty$.

Solution: First we show that for all $n \geq xy$, $n \in S$. It suffices to show that there exist nonnegative integers a, b such that $ax + by = n$. Since x and y are relatively prime, there exists b, $0 \leq b < x$, satisfying $by \equiv n \pmod{x}$. Now we can take $a = \frac{n-by}{x}$, which is positive by the assumption $n \geq xy$. Thus, $|T| < \infty$.

Sort the elements of T in increasing order so $t_1 < t_2 < \cdots < t_{|T|}$. Since $t_i \notin S$, at least one of m and $t_i - m$ is not in S for each $1 \leq m \leq \lfloor t_i/2 \rfloor$. Since there are only $i-1$ positive integers less than t_i and not in S, we have

$$\left\lfloor \frac{t_i}{2} \right\rfloor \leq i - 1,$$

or $t_i \leq 2i - 1$. Summing over i's gives

$$t_1 + t_2 + \cdots + t_{|T|} \leq |T|^2,$$

as desired.

20. In a forest each of 9 animals lives in its own cave, and there is exactly one separate path between any two of these caves. Before the election for Forest Gump, King of the Forest, some of the animals make an election campaign. Each campaign-making animal—$\mathcal{FGC}$ (Forest Gump candidate)—visits each of the other caves exactly once, uses only the paths for moving from cave to cave, never turns from one path to another between the caves,

and returns to its own cave at the end of the campaign. It is also known that no path between two caves is used by more than one $\mathcal{FGC}$. Find the maximum possible number of $\mathcal{FGC}$'s.

Solution: We translate this problem into the language of graph theory. Let each cave represent a vertex, and each path between a pair of caves represent the edge connecting the two vertices. We obtain a complete graph K_9. We are looking for the maximum number of Hamiltonian cycles without common edges in this complete graph K_9. (It is not hard to see that the number of Hamiltonian cycles is less than the number of vertices. Hence we can always pick a distinct $\mathcal{FGC}$ for each Hamiltonian cycle.)

The general result is that there are $\lfloor (n-1)/2 \rfloor$ disjoint Hamiltonian cycles in a complete graph K_n. Since there are $n(n-1)/2$ edges in K_n and each Hamiltonian cycle has n edges, there are at most $\lfloor (n-1)/2 \rfloor$ Hamiltonian cycles in K_n. We consider the following cases.

- *Case 1:* n is odd. We assume that $n = 2k+1$ for some positive integer k (as $n = 1$ is meaningless). We evenly arrange vertices $P_1, P_2, \ldots, P_{2k}$ around a circle in clockwise order and place vertex P_0 at the center of the circle. The first Hamiltonian cycle is

$$(P_0, P_1, P_2, P_{2k}, P_3, P_{2k-1}, P_4, P_{2k-2}, P_5, \ldots, \\ P_{k-1}, P_{k+3}, P_k, P_{k+2}, P_{k+1}, P_0).$$

  We can then rotate this cycle clockwise by angles of $\pi/k, 2\pi/k, \ldots, (k-1)\pi/k$ to obtain $k-1$ more cycles for a total of $k = \lfloor (n-1)/2 \rfloor$ cycles.

- *Case 2:* n is even. We assume that $n = 2k+2$ for some positive integer k. We evenly arrange vertices $P_1, P_2, \ldots, P_{2k}$ around a circle in clockwise order and place vertex P_0 at the center of the circle. We then place vertex P_{2k+1} somewhere inside the circle. For each Hamiltonian cycle in case 1, we put P_{2k+1} right in the middle of that path to obtain the $k = \lfloor (n-1)/2 \rfloor$ cycles.

For our problem, $n = 9$. Hence there are $\lfloor (n-1)/2 \rfloor = 4$ Hamiltonian cycles and thus a maximum of 4 $\mathcal{FGC}$'s.

21. [USA 1998, Franz Rothe] For a sequence $A_1, \ldots, A_n$ of subsets of $\{1, \ldots, n\}$ and a permutation π of $S = \{1, \ldots, n\}$, we define the diagonal set

$$D_\pi(A_1, A_2, \ldots, A_n) = \{i \in S \mid i \notin A_{\pi(i)}\}.$$

What is the maximum possible number of distinct sets which can occur as diagonal sets for a single choice of $A_1, \ldots, A_n$?

Solution: The answer is $2^n - n$.

We claim that $D_\pi(A_1, A_2, \ldots, A_n) \neq A_i$ for all i. It is clear that $D_{\pi_0}(A_1, A_2, \ldots, A_n) \neq A_i$, where $\pi_0(i) = i$. Hence

$$D_\pi(A_1, A_2, \ldots, A_n) = D_{\pi_0}(A_{\pi(1)}, A_{\pi(2)}, \ldots, A_{\pi(n)}) \neq A_{\pi(i)}.$$

But $\{A_{\pi(1)}, A_{\pi(2)}, \ldots, A_{\pi(n)}\} = \{A_1, A_2, \ldots, A_n\}$, from which our claim follows.

There are 2^n different subsets. Excluding the n original sets leaves at most $2^n - n$ possible sets as diagonal sets. Indeed, this number is obtainable. Let $A_i = \{i\}$ for $1 \leq i \leq n$. Then

$$D_{\pi_0}(A_1, A_2, \ldots, A_n) = \emptyset$$

and

$$D_\pi(A_1, A_2, \ldots, A_n) = \{i \in S \mid i \notin A_{\pi(i)}\} = \{i \in S \mid i \neq \pi(i)\}.$$

By appropriate choice of the permutation π, each subset of S with at least two elements can be created as a D_π. Hence the empty set and all subsets of S with at least two elements are exactly the possible diagonal sets, for a total of $2^n - n$ possibilities.

22. [IMO Shortlist 1994] A subset M of $\{1, 2, 3, \ldots, 15\}$ does not contain three elements whose product is a perfect square. Determine the maximum number of elements in M.

Solution: For a set M, let $|M|$ denote the number of elements in M. We say set M is *good* if it is a subset of $S = \{1, 2, 3, \ldots, 15\}$ and it does not contain three elements whose product is a perfect square. We want to find the maximum value of $|M|$, where M is good. Let m denote this maximum. We call a triple of numbers $\{i, j, k\}$, $1 \leq i < j < k \leq 15$ *bad* if ijk is a perfect square.

First we show that $m \leq 11$. Since there are disjoint bad triples $B_1 = \{1, 4, 9\}$, $B_2 = \{2, 6, 12\}$, $B_3 = \{3, 5, 15\}$, $B_4 = \{7, 8, 14\}$, if $|M| = 12$, all three numbers in at least one of these triples are in M. Hence M is not good if $|M| \geq 12$ and we conclude that $m \leq 11$.

If $m = 11$, then let M be a good set with $|M| = 11$. Then $M = S - \{a_1, a_2, a_3, a_4\}$, where $a_i \in B_i$ for $i = 1, 2, 3, 4$. Hence $10 \in M$. Since

$10 \in M$ and $B_1 = \{1, 4, 9\}$, $B_4 = \{7, 8, 14\}$, $B_5 = \{2, 5, 10\}$, $B_6 = \{6, 15, 10\}$ are bad triples with 10 as the only repeated element, $M = S - \{b_1, b_4, b_5, b_6\}$, where $b_1 \in B_1$, $b_4 \in B_4$, $b_5 \in \{2, 5\}$, and $b_6 \in \{6, 15\}$. Therefore $\{3, 12\} \subset M$. Then 1, 4, 9 are not in M. Since there are still two disjoint bad triples $\{2, 3, 6\}$ and $\{7, 8, 14\}$, we need to delete at least two more numbers to make M good. Hence $|M| \leq 10$, which contradicts the assumption that $|M| = 11$. Hence our assumption was wrong and $m \leq 10$.

It is not difficult to check that the set $\{1, 4, 5, 6, 7, 10, 11, 12, 13, 14\}$ satisfies the conditions of the problem. Hence 10 is the maximum number of elements in M.

23. [IMO Shortlist 2001] Find all finite sequences $(x_0, x_1, \ldots, x_n)$ such that for every $j, 0 \leq j \leq n, x_j$ equals the number of times j appears in the sequence.

Solution: Let $(x_0, x_1, \ldots, x_n)$ be any such sequence. Since each x_j is the number of times j appears, the terms of the sequence are nonnegative integers. Note that $x_0 > 0$ since $x_0 = 0$ is a contradiction. Let m denote the number of positive terms among $x_1, x_2, \ldots, x_n$. Since $x_0 = p \geq 1$ implies $x_p \geq 1$, we see that $m \geq 1$. Observe that $\sum_{i=1}^{n} x_i = m + 1$ since the sum on the left counts the total number of positive terms of the sequence, and $x_0 > 0$. (*Note:* For every $j > 0$ that appears as some x_i, the sequence is long enough to include a term x_j to count it, because the sequence contains j values of i and at least one other value, the value j itself if $i \neq j$ and the value 0 if $i = j$.) Since the sum has exactly m positive terms, $m - 1$ of its terms equal 1, one term equals 2, and the remainder are 0. Therefore only x_0 can exceed 2, so for $j > 2$ the possibility that $x_j > 0$ arises only in case $j = x_0$. In particular, $m \leq 3$. Hence there are three cases to consider. In each case, bear in mind that $m - 1$ of the terms $x_1, x_2, \ldots, x_n$ equal 1, one term equals 2, and the the others are 0.

(i) $m = 1$. We have $x_2 = 2$ since $x_1 = 2$ is impossible. Thus $x_0 = 2$ and the final sequence is $(2, 0, 2, 0)$.

(ii) $m = 2$. Either $x_1 = 2$ or $x_2 = 2$. The first possibility leads to $(1, 2, 1, 0)$ and the second one gives $(2, 1, 2, 0, 0)$.

(iii) $m = 3$. In this case, $x_p > 0$ for some $p \geq 3$. Thus $x_0 = p$ and $x_p = 1$. Then $x_1 = 1$ is contradictory, so $x_1 = 2$, $x_2 = 1$, and we have accounted for all of the positive terms of the sequence. The resulting sequence is $(p, 2, 1, \underbrace{0, \ldots, 0}_{p-3}, 1, 0, 0, 0)$.

In summary, there are three special solutions and one infinite family:

$$(2,0,2,0),\ (1,2,1,0),\ (2,1,2,0,0),\ (p,2,1,\underbrace{0,\ldots,0}_{p-3},1,0,0,0),$$

for $p \geq 3$.

Note: If one considers the null set to be a sequence, then it too is a solution.

An expanded version of the problem allows for infinite sequences, and such solutions exist. One simple construction starts with a finite solution $(x_0, x_1, \ldots, x_n)$, sets $x_{n+1} = n+1$ and continues as shown:

$$(x_0, x_1, \ldots, x_n, \underbrace{n+1, n+1, \ldots, n+1}_{x_{n+1}=n+1 \text{ terms}},$$
$$\underbrace{n+2, n+2, \ldots, n+2}_{x_{n+2} \text{ terms}}, \ldots).$$

For example,

$$(1,2,1,0,4,4,4,4,5,5,5,5,6,6,6,6,7,7,7,7,8,8,8,8,8,\ldots).$$

24. Determine if it is possible to partition the set of positive integers into sets $\mathcal{A}$ and $\mathcal{B}$ such that $\mathcal{A}$ does not contain any 3-element arithmetic sequence and $\mathcal{B}$ does not contain any infinite arithmetic sequence.

Solution: Each infinite arithmetic sequence is determined by its first term and its common difference, i.e., we can write an infinite arithmetic sequence $a, a+d, a+2d, \ldots$ as (a, d). Therefore we can define a bijection between the set of positive infinite arithmetic sequences and S, the set of all the lattice points in the first quadrant (not including the axes) of the coordinate plane. Note that the set S is countable as it can be counted by the sum of its coordinates:

$$\{(1,1);\ (1,2), (2,1);\ (1,3), (2,2), (3,1);\ \ldots\}.$$

We build set $\mathcal{A}$ inductively. At step 1, we put $a_1 = 1$ in set $\mathcal{A}$ (this breaks the infinite arithmetic sequence $(1,1)$); at step 2, we pick a number a_2 larger than $2a_1 = 2$ from the sequence $(1,2)$ and put it in $\mathcal{A}$ (this breaks the sequence $(1,2)$); at step 3, we pick a number a_3 that is larger than $2a_2$ from the sequence $(2,1)$ and put it in $\mathcal{A}$; ...; at the iþ step, $i \geq 3$, we pick a number that is larger than $2a_{i-1}$ from the iþ sequence from S (note that S

is countable so such an ordering exists) and put it in $\mathcal{A}$, and so on. All the numbers that are not in $\mathcal{A}$ form set $\mathcal{B}$.

By this construction, it is clear that every infinite sequence has been broken, so set $\mathcal{B}$ contains no infinite arithmetic sequence. On the other hand, the elements in $\mathcal{A}$ can be arranged in increasing order $a_1, a_2, a_3, \ldots$ with $a_{i+1} > 2a_i$. It follows that any three terms $a_i < a_j < a_k$ cannot form an arithmetic sequence as $2a_j < a_{j+1} \le a_k < a_k + a_i$.

Therefore it is possible to partition the set of positive integers into sets $\mathcal{A}$ and $\mathcal{B}$ such that $\mathcal{A}$ does not contain any 3-element arithmetic sequence and $\mathcal{B}$ does not contain any infinite arithmetic sequence.

25. [USSR 1989] Consider the set T_5 of 5-digit positive integers whose decimal representations are permutations of the digits 1, 2, 3, 4, 5. Determine if it is possible to partition T_5 into sets A and B such that the sum of the squares of the elements in A is equal to the corresponding sum for B.

Solution: We begin by making the following key observation.

Lemma. *Suppose $a_1, a_2, \ldots, a_5$ is a permutation of the digits 1, 2, 3, 4, 5. Then the sum of the squares of the 5-digit number $(a_1a_2a_3a_4a_5)$ and its four cyclic permutations is equal to the sum of the squares of the reverse number $(a_5a_4a_3a_2a_1)$ and its four cyclic permutations. That is, we have*

$$\sum_{i=1}^{5}(a_ia_{i+1}a_{i+2}a_{i+3}a_{i+4})^2 = \sum_{i=1}^{5}(a_{i+4}a_{i+3}a_{i+2}a_{i+1}a_i)^2,$$

where $a_{i+5} = a_i$.

Proof: We convert each decimal representation $(d_1d_2d_3d_4d_5)$ to $\sum 10^j d_j$ and expand the squares. The expansion creates square terms of the form $10^{2j}d_j^2$ and cross terms of the form $2 \cdot 10^{j+k}d_jd_k$. Now we consider the square terms and cross terms that occur on each side of the equality we want to prove. It is easy to see that the same square terms occur on both sides, since each digit occurs exactly once in each of the ones, tens, hundreds, thousands, and ten-thousands places. We claim that the two sides have identical cross terms, too. Indeed, each cross term $2 \cdot 10^{j+k}a_ja_k$ that arises from 10^ja_j and 10^ka_k on the left side also arises from 10^ka_j and 10^ja_k on the right, because the right side contains all of the reversed numbers. ■

We now divide the 120 numbers in T_5 into 24 groups that each contain 5 numbers that are cyclic permutations of each other. We put all permutations in which the numbers 1, 2, 3 occur cyclically in that order in A, and the rest (in which the order is 1, 3, 2) in B. Now each group of 5 numbers in A has a corresponding group in B that we can apply the lemma to, and this shows that A and B have equal sums of squares.

26. [China 1996] Let n be a positive integer. Find the number of polynomials $P(x)$ with coefficients in $\{0, 1, 2, 3\}$ such that $P(2) = n$.

First Solution: Let $S = \{0, 1, 2, 3\}$, and let

$$P(x) = a_m x^m + a_{m-1}x^{m-1} + \cdots + a_1 x + a_0,$$

where $a_i \in S$. Then $P(2) = 2^m a_m + 2^{m-1}a_{m-1} + \cdots + 2a_1 + a_0$. We are trying to find the number of sequences $(a_0, a_1, \ldots)$ with each $a_i \in S$ such that

$$a_0 + 2a_1 + 4a_2 + \cdots = \sum_{i=0}^{\infty} 2^i a_i = n.$$

We consider the generating function

$$\begin{aligned} f(x) &= (1 + x + x^2 + x^3)(1 + x^2 + x^4 + x^6) \\ &\quad (1 + x^4 + x^8 + x^{12}) \cdots, \end{aligned}$$

where $1+x+x^2+x^3$ represents the different choices for a_0, $1+x^2+x^4+x^6$ represents the different choices for a_1, $1 + x^4 + x^8 + x^{12}$ represents the different choices for a_2, and so on. It suffices to find the coefficient of term x^n in $f(x)$. Note that

$$\begin{aligned} f(x) &= \frac{x^4-1}{x-1} \cdot \frac{x^8-1}{x^2-1} \cdot \frac{x^{16}-1}{x^4-1} \cdot \frac{x^{64}-1}{x^8-1} \cdots \\ &= \frac{1}{(x-1)(x^2-1)}, \end{aligned}$$

as each term in the numerator occurs in the denominator of the fraction two terms away. By partial fractions, we obtain

$$\begin{aligned} f(x) &= \frac{1}{4(x+1)} - \frac{1}{4(x-1)} + \frac{1}{2(x-1)^2} \\ &= \frac{-2}{4(x^2-1)} + \frac{1}{2(x-1)^2} = \frac{1}{2}\left((x-1)^{-2} + \frac{1}{1-x^2}\right). \end{aligned}$$

Expanding the two functions in the last equation, we find that

$$\begin{aligned} f(x) &= \frac{1}{2}\left[\left(1 - \binom{-2}{1}x + \binom{-2}{2}x^2 - \cdots\right)\right. \\ &\quad \left. + \left(1 + x^2 + x^4 + \cdots\right)\right]. \end{aligned}$$

Since

$$\binom{-2}{n} = \frac{(-2)(-3)\cdots(-2-n+1)}{n!} = (-1)^n(n+1),$$

we obtain

$$\begin{aligned} f(x) &= \frac{1}{2}\Big[(1+2x+3x^2+\cdots)+(1+x^2+x^4+\cdots)\Big] \\ &= 1+x+2x^2+2x^3+3x^4+3x^5+\cdots \\ &= \sum_{m=0}^{\infty}\left(\left\lfloor\frac{m}{2}\right\rfloor+1\right)x^m. \end{aligned}$$

Thus, the coefficient of x^n is $\lfloor n/2\rfloor+1$, that is, there are $\lfloor n/2\rfloor+1$ polynomials satisfying the conditions of the problem.

Second Solution: We will solve a more general problem: Let m and n be positive integers with $m \geq 2$. Find the number of polynomials $P(x)$ with coefficients in $\{0, 1, 2, \ldots, m^2-1\}$ such that $P(m) = n$. We call such polynomials *good*.

Let $P(x) = \sum_{k=0}^{\infty} a_k x^k$, where $a_k \in \{0, 1, 2, \ldots, m^2-1\}$ for $k = 0, 1, 2, \ldots$. Then each a_k can be written in the form of $b_k m + c_k$, where $b_k, c_k \in \{0, 1, 2, \ldots, m-1\}$. Hence

$$n = P(m) = \sum_{k=0}^{\infty} b_k m^{k+1} + \sum_{k=0}^{\infty} c_k m^k = mt + \sum_{k=0}^{\infty} c_k m^k,$$

where $t = \sum_{k=0}^{\infty} b_k m^k$. For each t, $0 \leq t \leq \lfloor n/m\rfloor$, there is a unique way to write $t = \sum_{k=0}^{\infty} b_k m^k$ with $b_k \in \{0, 1, 2, \ldots, m-1\}$ (that is, express t in base m) and there is a unique way to write $n - mt = \sum_{k=0}^{\infty} c_k m^k$ with $c_k \in \{0, 1, 2, \ldots, m-1\}$ (that is, express $n - mt$ in base m). Hence we have a a bijection between the set $\{0, 1, \ldots, \lfloor n/m\rfloor\}$ and the set of good polynomials. Therefore there are $\lfloor n/m\rfloor + 1$ good polynomials.

For our problem, we have $m = 2$ and thus there are $\lfloor n/2\rfloor+1$ polynomials satisfying the conditions of the problem.

27. [IMO Shortlist 2000] Let n and k be positive integers such that $\frac{1}{2}n < k \leq \frac{2}{3}n$. Find the least number m for which it is possible to place each of m pawns on a square of an $n \times n$ chessboard so that no column or row contains a block of k adjacent unoccupied squares.

Solution: Call a placement of pawns on the board *good* if there is no $k \times 1$

(or $1 \times k$) block of unoccupied squares. Label the rows and the columns 0 through $n - 1$.

A standard good placement is obtained by putting pawns on squares (i, j) (iþ row and jþ column) such that $i + j + 1$ is divisible by k. Since $n < 2k$, the sum $i + j$ has to be equal to $k - 1$, $2k - 1$, or $3k - 1$, so the pattern is composed of at most three oblique lines. Since $3k \leq 2n$, these three lines consist of k squares, $2n - 2k$ squares, and $2n - 3k$ squares, respectively. This gives a total of $4(n - k)$ occupied squares.

We now show that this is actually the least possible number of pawns in a good placement. Suppose that we have a good placement of m pawns. Partition the board into nine rectangular regions

$$\begin{array}{ccc} A & B & C \\ D & E & F \\ G & H & I \end{array}$$

so that the corner regions A, C, G, I are $(n - k) \times (n - k)$ squares, each of regions B and H has $n - k$ rows and $2k - n$ columns, and each of regions D and F has $2k - n$ rows and $n - k$ columns. (This is possible since $2k - n > 0$.)

Observe that we can cut the region $A \cup B$ into $(n - k)$ $1 \times k$ rectangular strips. Similarly, we can obtain $n - k$ horizontal strips from $B \cup C$, $G \cup H$, and $H \cup I$. We can likewise obtain $4(n - k)$ vertical strips, for a total of $8(n - k)$ strips. By the conditions of the problem, each strip must contain at least one pawn. On the other hand, each pawn belongs to no more than two of the strips in our construction. Hence there are at least $4(n - k)$ pawns.

28. [China 1996] In a soccer tournament, each team plays another team exactly once and receives 3 points for a win, 1 point for a draw, and 0 points for a loss. After the tournament, it is observed that there is a team which has earned both the most total points and won the *fewest* games. Find the smallest number of teams in the tournament for which this is possible.

Solution: We call this special team W. Suppose that there are n teams in the tournament. There are $\binom{n}{2} = n(n - 1)/2$ games played for a total of at least $n(n - 1)$ points. Thus the average points earned per team is at least $n - 1$. Since W played $n - 1$ games and its score must be higher than the average, W won at least 1 game. Each of the other teams has to win at least 2 games for a total of at least 6 points; hence W has to tie at least 4 games (for a total of at least 7 points). But if team A tied its game against team W, team A has 7 points. Hence team W has to tie at least 5 games. Therefore, $n \geq 7$.

If $n = 7$, then team W won 1 game and tied 5 games for a total of 8 points. So each of the other teams won exactly 2 games and tied at most 1 game.

Hence each of the other teams must lose at least 3 games. Then there are at least $6 \times 3 = 18$ losses but only $1 + 6 \times 2 = 13$ wins, which is impossible. Hence $n \geq 8$.

We now give an example that shows that it is possible to have an 8-team tournament that satisfies the conditions of the problem. Let $W, A_1, A_2, \ldots, A_7$ be the 8 teams. Team W won games against A_1 and A_2, and tied the other games for a total of 11 points. For $1 \leq i \leq 7$, team A_i won its games against teams $A_{i+1}, A_{i+2}, A_{i+3}$ and lost its games against teams $A_{i+4}, A_{i+5}, A_{i+6}$, where $A_{i+7} = A_i$. Thus teams A_1 and A_2 each have 3 wins and 4 losses for a total of 9 points, and teams $A_3, A_4, \ldots A_7$ each have 3 wins, 3 losses, and 1 tie for a total of 10 points.

Thus the desired minimum is 8.

29. Let $a_1, \ldots, a_n$ be the first row of a triangular array with $a_i \in \{0, 1\}$. Fill in the second row $b_1, \ldots, b_{n-1}$ according to the rule $b_k = 1$ if $a_k \neq a_{k+1}$, $b_k = 0$ if $a_k = a_{k+1}$. Fill in the remaining rows similarly. Determine with proof the maximum possible number of 1's in the resulting array.

Solution: Let x_n denote the desired maximum number for an array with n rows. One can check that $x_1 = 1, x_2 = 2, x_3 = 4$.

Now we shall relate x_{n+3} to x_n. Consider the top three rows of an $(n+3)$-row triangle: $a_1, \ldots, a_{n+3}$; $b_1, \ldots, b_{n+2}$; $c_1, \ldots, c_{n+1}$. Get your pebbles ready. If at least one of a_k, b_k, c_k is 0, then place a pebble over column k corresponding to that 0. If $a_k b_k c_k = 1$, then $a_{k+1} = b_{k+1} = 0$; place a pebble over column k corresponding to a_{k+1} and another over column $k + 1$ corresponding to b_{k+1}. Starting with $k = 1$, run the above process repeatedly, each time letting k jump to the next pebbleless column. By the end, columns 1 through $n + 1$ will all have pebbles over them. If column $n + 2$ does not yet have a pebble over it, then none of the pebbles that have been placed corresponds to a_{k+2}, b_{k+2}, or a_{k+3}, but since there must be at least one 0 among those three numbers, we may place another pebble corresponding to it.

Since each of our $n + 2$ pebbles corresponds to a 0, there are at least $(n + 2)$ 0's in the top three rows of our triangle. Consequently, there can be at most $(2n + 4)$ 1's there, so $x_{n+3} \leq x_n + 2n + 4$. One can show by induction that

$$x_n \leq \left\lfloor \frac{n^2 + n + 1}{3} \right\rfloor.$$

Furthermore, this bound is attainable, as is demonstrated by the following pattern:

```
    1 1 0 1 1 0 1 1 0
     0 1 1 0 1 1 0 1
      1 0 1 1 0 1 1
     -------------
       1 1 0 1 1 0
        0 1 1 0 1
         1 0 1 1
        -------
          1 1 0
           0 1
            1
```

Each row's numbers repeat in blocks of three. From the bottom row up, the numbers of 1's in each row are 1, 1, 2, 3, 3, 4, 5, 5, 6, ...

30. There are 10 cities in the *Fatland*. Two airlines control all of the flights between the cities. Each pair of cities is connected by exactly one flight (in both directions). Prove that one airline can provide two traveling cycles with each cycle passing through an odd number of cities and with no common cities shared by the two cycles.

Solution: Let each city be a vertex, and each flight between each pair of cities an edge between the corresponding vertices. We color an edge blue if it is from one airline and red otherwise. This gives us a 2-colored complete graph K_{10}. In the language of graph theory, we must show that in a 2-colored complete graph K_{10} there are two monochromatic non-intersecting odd cycles. We start with a well-known result in graph theory.

Lemma 1. *If the edges of a complete graph K_6 are colored in 2 colors, the graph contains a monochromatic triangle.*

Proof: There is a proof using simple arguments involving the Pigeonhole Principle. But we present a *cool* proof. We show that indeed there are two monochromatic triangles. Let $v_1, v_2, \ldots, v_6$ be the vertices of K_6. If a pair of edges v_iv_j and v_iv_k are of the same color, then we call angle $v_jv_iv_k$ *monochromatic*. Let r_i and b_i be the respective numbers of red and blue edges emanating from v_i. Then $r_i + b_i = 5$ for all i and there are

$$\sum_{i=1}^{6}\left(\binom{r_i}{2}+\binom{b_i}{2}\right) \geq \sum_{i=1}^{6}\left(\binom{2}{2}+\binom{3}{2}\right) = 24$$

monochromatic angles. On the other hand, in each monochromatic triangle, there are three monochromatic angles, while in each of the other triangles, there is one monochromatic angle. Let m be the number of monochromatic triangles. Since there are a total of $\binom{6}{3} = 20$ triangles, there are $3m + (20 - m) = 20 + 2m$ monochromatic angles. Therefore, $20 + 2m \geq 24$ or $m \geq 2$, as desired. ■

Lemma 2. *If the edges of a complete graph K_5 are colored in 2 colors and the graph does not contain a monochromatic triangle, then it consists of two length-5 monochromatic cycles.*

Proof: Let $v_1, v_2, \ldots, v_5$ be the vertices of K_5. If three of the edges v_1v_2, $v_1v_3, \ldots, v_1v_5$ are the same color, then we have a monochromatic triangle. Indeed, we may assume that v_1v_2, v_1v_3, v_1v_4 are red; then if any of v_2v_3, v_3v_4, v_4v_2 are red, we are done. Otherwise, $v_2v_3v_4$ is a blue triangle, and we are also done. (This argument can be easily extended to prove the existence of a monochromatic triangle in a 2-colored K_6.) Since there are no monochromatic triangles in our K_5, there are two red edges and two blue edges from each vertex. If we only look at the red edges, we have a subgraph of 5 vertices with each vertex having degree 2. Hence this subgraph is either a cycle or can be decomposed into a few non-intersecting cycles. But since there are only 5 vertices, it cannot have two cycles. Hence we must have a red cycle of length 5. In exactly the same way, we can prove that we must have a blue cycle of length 5. ■

Now we are ready to prove our main result. Let $v_1, v_2, \ldots, v_{10}$ be the vertices of our 2-colored (red and blue) complete graph K_{10}. By Lemma 1, there is a monochromatic triangle in K_{10}. Without loss of generality, say it is $v_1v_2v_3$. By Lemma 1 again, there is a monochromatic triangle in the subgraph $K_{10} - \{v_1, v_2, v_3\}$. Without loss of generality, assume it is $v_4v_5v_6$. If $v_1v_2v_3$ and $v_4v_5v_6$ are of the same color, we are done. If not, assume that $v_1v_2v_3$ is blue and $v_4v_5v_6$ is red. Consider the edges v_iv_j, $1 \leq i \leq 3$ and $4 \leq j \leq 6$. By the Pigeonhole Principle, five of them are of the same color; without loss of generality, assume they are blue. Hence there is some j_0, $4 \leq j_0 \leq 6$, such that two of the edges $v_{j_0}v_1$, $v_{j_0}v_2$, $v_{j_0}v_3$ are blue. Therefore we have one blue triangle and a red triangle with exactly one common vertex v_{j_0}.

For simplicity, we relabel the points so that $v_1v_2v_3$ is blue and $v_3v_4v_5$ is red. Consider the subgraph $K_{10} - \{v_1, v_2, \ldots, v_5\}$. If it has a monochromatic triangle, we are done as we can pick one of the triangles $v_1v_2v_3$ and $v_3v_4v_5$ to match the color of this new triangle. Therefore one airline can provide two traveling cycles of three cities with no cities in common. If not, by Lemma 2, we have a red length-5 cycle and a blue length-5 cycle. Therefore each airline can provide a traveling cycle of three cities and traveling cycle of five cities with no common cities.

31. [MOSP 1997] Suppose that each positive integer not greater than $n(n^2-2n+3)/2$, $n \geq 2$, is colored one of two colors (red or blue). Show that there must be a monochromatic n-term sequence $a_1 < a_2 < \cdots < a_n$ satisfying

$$a_2 - a_1 \leq a_3 - a_2 \leq \cdots \leq a_n - a_{n-1}.$$

Solution: Call a sequence $a_1, a_2, \ldots, a_n$ such that

$$a_2 - a_1 \leq a_3 - a_2 \leq \cdots \leq a_n - a_{n-1} \leq m$$

an n-term m-sequence. Note that

$$s_n = \frac{n(n^2-2n+3)}{2} = 3\binom{n}{3} + \binom{n}{2} + \binom{n}{1}.$$

We shall prove that if the integers are colored red and blue, the first s_n integers contain a monochromatic n-term $3\binom{n}{2}$-sequence. We induct on n. The base case $n = 2$ is trivial.

Assume without loss of generality that there is a red n-term $3\binom{n}{2}$-sequence $a_1, a_2, \ldots, a_n$ with $a_n \leq s_n$. Note that

$$\begin{aligned} & s_{n+1} - s_n \\ = & \left[3\binom{n+1}{3} + \binom{n+1}{2} + \binom{n+1}{1}\right] - \left[3\binom{n}{3} + \binom{n}{2} + \binom{n}{1}\right] \\ = & 3\binom{n}{2} + \binom{n}{1} + 1. \end{aligned}$$

Consider the list of $n+1$ numbers

$$\begin{aligned} & a_n + 3\binom{n}{2}, a_n + 3\binom{n}{2} + 1, \ldots, a_n + 3\binom{n}{2} + n \\ < & s_n + 3\binom{n}{2} + \binom{n}{1} + 1 = s_{n+1}. \end{aligned}$$

If all of them are blue, then we have a blue $(n+1)$-term 1-sequence and we are done. Otherwise, one of them, say $(a + 3\binom{n}{2} + k)$, $0 \leq k \leq n$, is red. Let $a_{n+1} = a + 3\binom{n}{2} + k$. Then

$$a_{n+1} - a_n = 3\binom{n}{2} + k = 3\binom{n+1}{2} - 3\binom{n}{1} + k \leq 3\binom{n+1}{2},$$

and again we have a $n+1$-term $3\binom{n+1}{2}$ sequence. This completes the induction and our proof.

32. [C. J. Smyth] The set $\{1, 2, \ldots, 3n\}$ is partitioned into three sets A, B, and C with each set containing n numbers. Determine with proof if it is always possible to choose one number out of each set so that one of these numbers is the sum of the other two.

Solution: (V. Alexeev) Suppose that $\{1, 2, \ldots, 3n\}$ is partitioned into three sets A, B, and C, each set containing n numbers. For brevity, we shall call a triple (a, b, c) *good* if $a \in A, b \in B, c \in C$ and one of the numbers a, b, c is the sum of the remaining two.

Without loss of generality, we may assume that $1 \in A$ and, if k is the smallest number not in A, that $k \in B$. Assuming that there are no good triples, we claim

$$\text{If } x \in C, \text{ then } x - 1 \in A. \tag{2}$$

Then a contradiction will follow immediately from (2). Indeed, if $C = \{c_1, c_2, \ldots, c_n\}$, then A contains the numbers $c_1 - 1, c_2 - 1, \ldots, c_n - 1$, all of which are greater than 1 because $2 \notin C$. But $1 \in A$, so A would have at least $n + 1$ elements.

Now we prove our claim. Assume it is not true. Then there is a number $x \in C$ such that $x - 1 \notin A$. Clearly $x - 1 \notin B$ since otherwise $(1, x - 1, x)$ is good. Now, based on the fact that $x \in C$ and $x - 1 \in C$, we will prove that $x - k \in C$ and $x - k - 1 \in C$. (Recall that k is the smallest element not in A.) Indeed, if $x - k \in A$, then $(x - k, k, x)$ is good; if $x - k \in B$, then $(k - 1, x - k, x - 1)$ is good. Similarly, the relations $x - k - 1 \in A$ and $x - k - 1 \in B$ yield the good triples $(x - k - 1, k, x - 1)$ and $(1, x - k - 1, x - k)$, respectively. We can repeat this argument, concluding that all numbers $x - ik$ and $x - ik - 1$ are in C, for $i = 0, 1, 2, \ldots$, provided that they are positive. But $x - ik$ must be one of the numbers $1, 2, \ldots, k$ for some i. Hence it will be an element of either A or B, a contradiction. Therefore $x - 1 \in A$, proving (1), and we are done.

33. [MOSP 2002] Assume that each of the 30 MOPpers has exactly one favorite chess variant and exactly one favorite classical inequality. Each MOPper lists this information on a survey. Among the survey responses, there are exactly 20 different favorite chess variants and exactly 10 different favorite inequalities. Let n be the number of MOPpers M such that the number of MOPpers who listed M's favorite inequality is greater than the number of MOPpers who listed M's favorite chess variant. Prove that $n \geq 11$.

Solution: Let $c_1, c_2, \ldots, c_{20}$ denote the 20 different favorite chess variants, and let $e_1, e_2, \ldots, e_{10}$ denote the 10 different favorite inequalities. Let set

S_i, $1 \le i \le 20$, denote the set of all MOPpers who chose c_i as their favorite chess variant, and let set T_j, $1 \le j \le 10$, denote the set of all MOPpers who chose e_j as their favorite inequality.

For a set X, let $|X|$ denote the number of elements in X. Each MOPper M is assigned a pair of values (x_M, y_M) as follows: if $M \in S_i$, then $x_M = 1/|S_i|$, and if $M \in T_j$, then $y_M = 1/|T_j|$. We call this pair of values the coordinates of M. We are looking for all the MOPpers M such that $x_M > y_M$.

Summing up all the x-coordinates over all MOPpers yields 20, and summing up all the y-coordinates over all MOPpers yields 10. Hence

$$\sum_M (x_M - y_M) = 10.$$

Note that for each M, $x_M - y_M < x_M \le 1$. Therefore there are at least 11 terms in the above summation that are positive, i.e., there are at least 11 MOPpers M with $x_M > y_M$, which is what we wanted to prove.

34. [USAMO 1999 submission, Bjorn Poonen] Starting from a triple (a, b, c) of nonnegative integers, a *move* consists of choosing two of them, say x and y, and replacing one of them by either $x + y$ or $|x - y|$. For example, one can go from $(3, 5, 7)$ to $(3, 5, 4)$ in one move. Prove that there exists a constant $r > 0$ such that whenever a, b, c, n are positive integers with $a, b, c < 2^n$, there is a sequence of at most rn moves transforming (a, b, c) into (a', b', c') with $a'b'c' = 0$.

Solution: We will use strong induction on n to show that $r = 12$ works. The base case is trivial, as is the case $abc = 0$. For the induction step, we assume without loss of generality that $a \le b \le c$. Using two moves if necessary to replace a by $c - a$ and b by $c - b$, we may instead assume that $1 \le a \le b \le c/2$. Let m be the integer such that $2^{m-1} \le b < 2^m$. Since $1 \le b \le c/2 < 2^{n-1}$, we have $1 \le m \le n - 1$. Define a sequence $x_0 = a$, $x_1 = b$, and $x_k = x_{k-1} + x_{k-2}$ for $k \ge 2$.

Lemma. *Every integer $y \ge b$ can be expressed in the form*

$$\epsilon + x_{i_i} + \cdots + x_{i_\ell}$$

where $0 \le \epsilon < b$, $i_1 < i_2 < \cdots < i_\ell$, and $x_{i_\ell} \le y < x_{i_\ell+1}$.

Proof: Since x_i are increasing, there is a unique $i \ge 1$ for which $x_i \le y < x_{i+1}$. We use strong induction on i. If $y - x_i < b$, we let $\epsilon = y - x_i$ and we are done. Otherwise $x_1 = b \le y - x_i < x_{i+1} - x_i = x_{i-1}$. Thus there is a unique $j \ge 1$ such that $x_i \le y - x_i < x_{j+1}$, and $j < i$, so we finish by applying the inductive hypothesis to $y - x_i$. ■

Write $c = \epsilon + x_{i_1} + \cdots + x_{i_\ell}$, where $0 \leq \epsilon < b$ and $0 < i_1 < \cdots < i_\ell$. Since $x_{k+2} = x_{k+1} + x_k = 2x_k + x_{k-1} \geq 2x_k$ for $k \geq 1$, we have

$$x_{2n-2m+3} \geq 2^{n-m+1}x_1 \geq 2^{n-m+1}2^{m-1} = 2^n > c,$$

so $i_1 < i_2 < \cdots < i_\ell < 2n - 2m + 3$.

Using $2n - 2m + 1$ addition moves we can change $(a, b, c) = (x_0, x_1, c)$ into (x_2, x_1, c), then into (x_2, x_3, c), and so on, until we reach the triple $(x_{2n-2m+2}, x_{2n-2m+1}, c)$. Along the way, we intersperse at most $2n-2m+2$ moves between these, to subtract from c the x_{i_j} in the representation of c as they are produced in the first two coordinates. Thus we will eventually reduce c to ϵ. Now we can perform $2n-2m+1$ subtraction moves to change the triple $(x_{2n-2m+2}, x_{2n-2m+1}, \epsilon)$ back to the triple $(x_{2n-2m}, x_{2n-2m+1}, \epsilon)$, and so on, undoing the previous operations on the first two coordinates, until we end up with the triple (a, b, ϵ).

Reaching (a, b, ϵ) required at most

$$2 + (2n-2m+1) + (2n-2m+2) + (2n-2m+1) = 6n - 6m + 6$$

moves. Afterward, since $a, b, \epsilon < 2^m$, we can transform (a, b, ϵ) into a triple with a zero in at most $12m$ more moves, by the inductive hypothesis. Thus we have a total of at most $(6n - 6m + 6) + 12m = 6n + 6m + 6 \leq 12n$ moves, since $m \leq n - 1$.

35. [IMO Shortlist 1998] A rectangular array of numbers is given. In each row and each column, the sum of all the numbers is an integer. Prove that each nonintegral number x in the array can be changed into either $\lceil x \rceil$ or $\lfloor x \rfloor$ so that the row-sums and the column-sums remain unchanged. (Note that $\lceil x \rceil$ is the least integer greater than or equal to x, while $\lfloor x \rfloor$ is the greatest integer less than or equal to x.)

Solution: First, we replace all entries by $\lfloor x \rfloor$, and mark each change with a $-$ sign (so if x is an integer, there is no mark). Then we restore the column-sums by changing, column by column, the roundings of certain numbers, chosen arbitrarily, and changing their markings to $+$'s.

We then restore the row-sums without disturbing the column-sums. Denote by s the sum of the absolute values of the changes in the row-sums. It is necessarily an even integer (as the sum of all the numbers is preserved), and we want it to be 0. If $s > 0$, we will decrease it by 2 at a time.

We say row S is *accessible* from row R if there exists a column C such that $R \cap C$ is marked with a $+$ and $S \cap C$ is marked with a $-$. We may assume that the first row-sum is too high. Then it contains a +. Since column-sums are restored, there is a − in the same column as the + (otherwise all of the marks in the column would be +'s, and the column-sum would be too high). We may assume that the second row is accessible from the first row. If its row-sum is too low, we interchange the + and the − on the access-column along with the roundings which define the markings. This will decrease s by 2. If the second-row sum is not too low, then it must contain a +, and some row is accessible from it. If we eventually reach a row whose sum is too low, a chain of interchanges along the access-columns will decrease s by 2. We claim this happens.

Denote by A the union of all rows accessible from the first row, directly or indirectly, and including itself. Denote by B the union of all other rows. Let C be any column. If $A \cap C$ contains no +'s, then the sum of its entries has not increased from its original value. If $A \cap C$ contains at least one +, then $B \cap C$ contains no −'s, as otherwise some rows in B would have been accessible and should belong to A. Hence the sum of the entries of $B \cap C$ has not decreased, so that the sum of the entries of $A \cap C$ has not increased in this case, as well. Since C is arbitrary, we conclude that the sum of the entries of A has not increased. Since the first row-sum is too high, some row sum in A must be too low, justifying the claim.

36. [USAMO 1997 submission] A finite set of (distinct) positive integers is called a *DS-set* if each of the integers divides the sum of them all. Prove that every finite set of positive integers is a subset of some DS-set.

First Solution: We induct on the number of terms in our set that do not divide the sum of the elements of the set. For 0 terms, it is obvious that we have a DS-set.

Let Σ_S be the sum of the elements of set S and let $s \in S$ be such that $s \nmid \Sigma_S$. Suppose that $s = 2^k m$, where $2 \nmid m$.

(i) *Step 1:* Add the following elements to S: $\Sigma_S, 2\Sigma_S, 4\Sigma_S, \ldots, 2^{k-1}\Sigma_S$. The new set T has sum $\Sigma_T = 2^k \Sigma_S$. Hence all of the new elements divide Σ_T, as do all of the old elements that divided Σ_S. Note also that the elements of T are still distinct.

(ii) *Step 2:* If $m = 1$ we may skip this step. Otherwise, recall that by Euler's extension of Fermat's Little Theorem, $2^{\phi(m)} \equiv 1 \pmod{m}$. Let $r = \phi(m)$. Now we add to T the elements $2\Sigma_T, 4\Sigma_T, \ldots, 2^{r-1}\Sigma_T$, as well as $(2^r-1)\Sigma_T, 2(2^r-1)\Sigma_T, 4(2^r-1)\Sigma_T, \ldots, 2^{r-2}(2^r-1)\Sigma_T$. The new set U thus formed has sum $\Sigma_U = 2^{r-1}(2^r-1)\Sigma_T$. Thus all of the elements we added divide Σ_U, as do all of the elements of T that divided Σ_T. Furthermore, $m \mid \Sigma_U$.

After performing these two steps, we have $s \mid \Sigma_U$. Furthermore, all of the elements that originally divided Σ_S still divide Σ_U, and all of the elements that we added also divide Σ_U. Hence if we had n elements of S which did not divide Σ_S, we now have at most $n-1$ elements of T which do not divide Σ_T.

Thus, if we can construct a DS-set containing any subset with n elements not dividing the sum, then we can construct a DS-set containing as a subset any set with $n+1$ elements not dividing the sum. By induction, we are done.

Second Solution: (By Po-Ru Loh) For any $n \geq 2$, we exhibit a DS-set that contains the numbers $1, 2, \ldots, n$. Since for any finite set S of positive integers we can choose n large enough that $S \subseteq \{1, 2, \ldots, n\}$, this will suffice. First we put in our set the numbers $1, 2, \ldots, n$ and $n(n+1)/2$. This will bring the sum up to $n(n+1)$. Now we add the numbers $(n-j)(n-j+2)(n-j+3)\cdots(n)(n+1)$ for $j = 2, 3, \ldots, n-1$. Thus the overall sum is

$$
\begin{aligned}
n(n+1) &+ \sum_{j=2}^{n-1}(n-j)(n-j+2)\cdots(n+1) \\
&= n(n+1) + \sum_{j=2}^{n-1}[(n-j+1)(n-j+2)\cdots(n+1) - \\
&\qquad\qquad (n-j+2)(n-j+3)\cdots(n+1)] \\
&= n(n+1) + (n+1)! - n(n+1) \\
&= (n+1)!,
\end{aligned}
$$

since the sum telescopes. It is clear that the elements of our set are distinct and divide $(n+1)!$, so the proof is complete.

37. [China 1994, Chengzhang Li] Twelve musicians $M_1, M_2, \ldots, M_{12}$ gather at a week-long chamber music festival. Each day, there is one scheduled concert and some of the musicians play while the others listen as members of the audience. For $i = 1, 2, \ldots, 12$, let t_i be the number of concerts in which musician M_i plays, and let $t = t_1 + t_2 + \cdots + t_{12}$. Determine the minimum value of t such that it is possible for each musician to listen, as a member of the audience, to all the other musicians.

Solution: The condition of the problem is the following:

If a musician is not performing on a given day, he observes the concert as a member of the audience. If a musician is performing on a given day, he cannot observe other musicians' performances on that day. Each musician is required to observe at least one of each of the other musicians' performances. $(*)$

- *Observation 1.* To achieve $(*)$, any three musicians must perform in at least 3 concerts. Indeed, if they only perform in 2 concerts, by the Pigeonhole Principle, two of them perform in 1 concert. So they cannot observe each other on that day. This means they have to observe each other in the other concert, which is impossible.
- *Observation 2.* To achieve $(*)$, any 7 or more musicians must perform in at least 4 concerts. Indeed, if they only perform in 3 concerts, by the Pigeonhole Principle, there are at least 3 musicians performing in 1 concert, so they cannot observe each other in that concert. Then they have to observe each other in the other 2 concerts, which is impossible by observation 1.
- *Observation 3.* To achieve $(*)$, any 9 musicians must perform in at least 5 concerts. Indeed, if they only perform in 4 concerts, then each of them can perform in at most 3 concerts, as otherwise he cannot listen to the other 8 musicians. Note that if one of them only performs in 1 concert, then all 8 other musicians must observe that concert. Then these 8 musicians only have 3 concerts to observe each other, which is impossible by observation 2. Also note that if one of them performs in 3 concerts, then he can only listen in the fourth concert; hence all the other 8 musicians must perform in that concert. This again leads to the situation that the other 8 musicians have to observe each other in 3 concerts, which is impossible by observation 2. Therefore each of the 9 musicians performs in 2 concerts. There are $\binom{4}{2} = 6$ ways to choose 2 concerts to perform. By the Pigeonhole Principle, there are two musicians who have the same performing dates so they cannot observe each other, which violates $(*)$.

We assume that there are k musicians who each perform in only 1 concert. These k musicians must perform in different concerts as otherwise they cannot observe each other. Hence $0 \le k \le 7$. Note that all of these k concerts must be solo concerts. The remaining $12 - k$ musicians each perform in at least 2 concerts, and they must observe each other in the $7 - k$ concerts left. It is easy to see that it is impossible for $k = 7$ or 6. If $k = 5$, 7 musicians must observe each other in 2 concerts, which is impossible by observation 2; if $k = 4$, 8 musicians must observe each other in 3 concerts, which is impossible by observation 2; and if $k = 3$, 9 musicians must observe each other in 4 concerts, which is impossible by observation 3. Hence $k \le 2$, so $t \ge k + 2(12 - k) \ge 22$.

Finally we give an example that shows that $t = 22$ is indeed achievable. Let musicians M_1 and M_2 give solo performances on days 1 and 2. Each of the other 10 musicians will perform twice. There are 5 remaining days and hence $\binom{5}{2} = 10$ ways to select two days on which to perform. Thus letting each musician perform on a different pair of days completes the example.

38. [USAMO 1999 submission, Richard Stong] An $m \times n$ array is filled with the numbers $\{1, 2, \ldots n\}$, each used exactly m times. Show that one can always permute the numbers within columns to arrange that each row contains every number $\{1, 2, \ldots, n\}$ exactly once.

Solution: It suffices to show that we can permute the numbers within columns to arrange that the top row contains every number $1, 2, \ldots, n$ exactly once; the result then follows by induction on the number of rows. For this we apply the Marriage Lemma. For the *boys* take the columns, and for the *girls* take the numbers $\{1, 2, \ldots, n\}$. Say a boy (column) likes a girl (number) if that number occurs in the column. For each set of k columns, there are a total of km numbers in those columns. Therefore there must be at least k different numbers among them. Thus there is a marriage of the columns and the numbers they contain. Permuting these numbers to the tops of their respective columns makes the first row have all n numbers.

39. [IMO Shortlist 1998] Let us set $U = \{1, 2, \ldots, n\}$, where $n \geq 3$. A subset S of U is said to be *split* by an arrangement of the elements of U if an element not in S occurs in the arrangement somewhere between two elements of S. For example, 13542 splits $\{1, 2, 3\}$ but not $\{3, 4, 5\}$. Prove that for any $n-2$ subsets of U, each containing at least 2 and at most $n-1$ elements, there is an arrangement of elements of U which splits all of them.

Solution: We induct on n. For $n = 3$, the family consists of a single 2-element subset $\{i, j\}$, which is split by the permutation (i, k, j), where k is the third element of U.

We now assume that the result holds for some positive integer $n \geq 3$, and let $U = \{1, 2, \ldots, n+1\}$. We are given a family $\mathcal{F}$ of $n-1$ subsets, each containing at least 2 and at most n elements. We have the following key observation.

Lemma. *There is an element of U which is contained in all n-element subsets of $\mathcal{F}$, but in at most one of its 2-element subsets.*

Proof: Suppose that $\mathcal{F}$ contains k 2-element subsets and ℓ n-element subsets. Then $k + \ell \leq n-1$. At most k elements of U can appear two or more times in the 2-element subsets. Hence the number of elements which appear at most once among them is at least $(n+1)-k \geq (n+1)-(n-1-\ell) = \ell+2$. Since there are only ℓ elements which are not contained in some of the ℓ n-element subsets, one of these $\ell + 2$ elements will satisfy the desired conditions. ■

Now we prove our main result. Without loss of generality, we may assume the number $n + 1$ is an element of U satisfying the property of the lemma. When it is removed, all n-element subsets in $\mathcal{F}$ become $(n - 1)$-element subsets of $\{1, 2, \ldots, n\}$, while at most one of the 2-element subsets of $\mathcal{F}$ becomes a singleton.

If we have such a subset $\{i\}$, then the induction hypothesis guarantees the existence of a permutation π of $\{1, 2, \ldots n\}$ that splits all the other $n - 2$ subsets (with $n + 1$ taken out). By adding $n + 1$ to π anywhere away from i, we have a permutation which splits all $n - 1$ subsets in $\mathcal{F}$. (Note that all of the other subsets that contain $n + 1$ which were already split before adding $n + 1$ remain split.)

If we do not have such a singleton subset, choose any subset S among the $n - 1$ subsets. By the induction hypothesis, we have a permutation π of $\{1, 2, \ldots, n\}$ which splits all of the other $n - 2$ subsets. If $n + 1 \notin S$, by adding $n + 1$ to π between two elements of S, we have a permutation which splits all $n - 1$ subsets in $\mathcal{F}$. Otherwise, if $n + 1 \in S$, if π does not already split S, we may add $n + 1$ on either the left or right end of π to split S. This completes our induction.

Note: If $\mathcal{F}$ contains $n - 2$ subsets each of which contains at least 3 and at most $n - 2$ elements, we have a much simpler approach. For any k-element subset S of U, there are $k!(n - k + 1)!$ permutations of U which do not split S; there are $k!$ ways to permute the elements of S and there are $(n - k + 1)!$ ways to permute the $n - k$ elements not in S and one big block of all the elements in S. The maximum value of $k!(n - k + 1)$, $3 \leq k \leq n - 2$, is $3!(n - 2)!$. So the total number of permutations which do not split some of the subsets in $\mathcal{F}$ is at most $(n - 2)3!(n - 2)!$, which is less than $n!$, the total number of all permutations. Hence there is a permutation that splits all subsets in $\mathcal{F}$.

40. [New York State Math League 2001/IMO Shortlist 2001] A pile of n pebbles is placed in a vertical column. This configuration is modified according to the following rules. A pebble can be moved if it is at the top of a column which contains at least two more pebbles than the column immediately to its right. (If there are no pebbles to the right, think of this as a column with 0 pebbles.) At each stage, choose a pebble from among those that can be moved (if there are any) and place it at the top of the column to its right. If

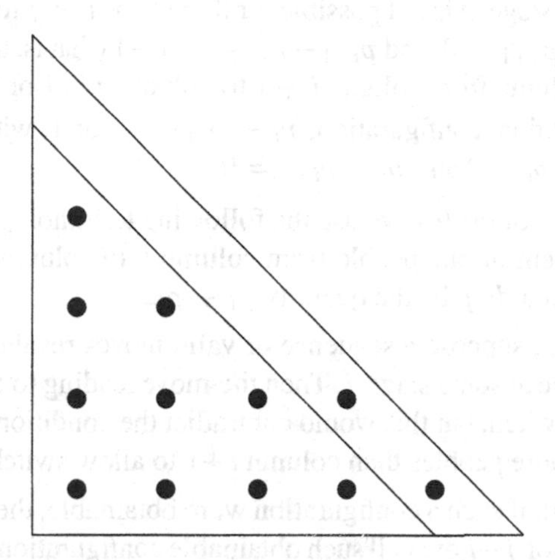

Final Configuration for $n = 12$

no pebbles can be moved, the configuration is called a *final configuration*. For each n, show that no matter what choices are made at each stage, the final configuration obtained is unique. Describe that configuration in terms of n.

First Solution: At any stage, let p_i be the number of pebbles in column i for $i = 1, 2, \ldots$, where column 1 denotes the leftmost column. We will show that in the final configuration, for all i for which $p_i > 0$ we have $p_i = p_{i+1} + 1$, except that for at most one i^*, $p_{i^*} = p_{i^*+1}$. Therefore, the configuration looks like the figure shown below, where there are c nonempty columns and there are from 1 to c pebbles in the last diagonal row of the triangular configuration. In particular, let $t_k = 1 + 2 + \cdots + k = k(k+1)/2$ be the kth triangular number. Then c is the unique integer for which $t_{c-1} < n \le t_c$. Let $s = n - t_{c-1}$. Then there are s pebbles in the rightmost diagonal, and so the two columns with the same height are columns $c - s$ and $c-s+1$ (except if $s = c$, in which case no nonempty columns have equal height).

Another way to say this is

$$p_i = \begin{cases} c - i & \text{if } i \le c - s, \\ c - i + 1 & \text{if } i > c - s. \end{cases}$$

To prove this claim, we show that

(a) At any stage of the process, $p_1 \ge p_2 \ge \cdots$.

(b) At any stage, it is not possible for there to be $i < j$ for which $p_i = p_{i+1}$, $p_j = p_{j+1} > 0$, and $p_{i+1} - p_j \le j - i - 1$ (that is, the average decrease per column from column $i+1$ to column j is 1 or less).

(c) At any final configuration, $p_i - p_{i+1} = 0$ or 1, with at most one i for which $p_i > 0$ and $p_i - p_{i+1} = 0$.

In the proofs of (a)-(c), we use the following terminology. Let a *k-switch* be the movement of one pebble from column k to column $k+1$, and for any column i let a *drop* be the quantity $p_i - p_{i+1}$.

To prove (a), suppose a sequence of valid moves resulted in $p_i < p_{i+1}$ for the first time at some stage j. Then the move leading to this stage must have been an i-switch, but this would contradict the condition that column i have at least 2 more pebbles than column $i+1$ to allow switches.

To prove (b), if such a configuration were obtainable, there would be a minimum value of $j-i$ over all such obtainable configurations, and we now show that there is no minimum. Suppose $p_1, p_2, \ldots$ was such a minimal configuration. It cannot be that $j = i+1$, for what would columns $i, i+1, i+2$ look like just before the move that made the heights equal? The move must have been a k-switch for $i - 1 \le k \le i + 2$, but if so the configuration before the switch was impossible (not decreasing).

Now suppose $j > i + 1$. Consider the first configuration C in the sequence for which columns $i, i+1, j, j+1$ are at their final heights. Note that from p_{i+1} to p_j the columns decrease by exactly one each time in C, because if there was a drop of 2 or more at some point, there would have to be another drop of 0 in this interval to obtain an average of 1 or less, and thus $j - i$ is not minimal. The move leading to C was either an i-switch or a j-switch. If it was the former, at the previous stage columns $i+1$ and $i+2$ had the same height, violating the minimality of $j-i$. A similar contradiction arises if the move was a j-switch.

Finally, to prove (c), if any drop is 2 or more, the configuration is not final. However, if all drops are 0 or 1, and there were two drops of 0 between nonempty columns (say between i and $i+1$ and between j and $j+1$), then (b) would be violated. Thus a final configuration that satisfies (b) also satisfies (c). It now follows easily that the only possible final configuration is the one described earlier.

Second Solution: At each stage, let c be the rightmost nonempty column. In conditions (a)–(c) in the previous solution, replace (b) by (b′), where

(b′) All configurations obtainable from the initial configuration satisfy

$$p_i - p_j \ge j - i - 1 \quad \text{for all } i < j \le c + 1. \tag{3}$$

(The restriction to $j \leq c+1$, which causes certain complications, is necessary for (3) to be true.) Fact (c), and thus the answer, follows as easily from (b′) as from (b). We prove (b′) by induction as follows.

Condition (3) is immediate for the initial configuration: Since $c = 1$, the only case is $p_1 - p_2 = n > 2-1-1$. Now suppose some configuration $p_1, p_2, \ldots$ with final nonempty column c_p satisfies (3), and a new configuration $q_1, q_2, \ldots$ is obtained from it by a k-switch. Thus $q_k = p_k - 1$, $q_{k+1} = p_{k+1} + 1$, and $q_i = p_i$ for all other i. Let the new configuration have c_q nonempty columns. Note that $c_q = c_p$ unless $k = c_p$.

For any $i < j \leq c_q + 1$ we now show that $q_i - q_j \geq j - i - 1$. The only cases to consider are those where $q_i - q_j < p_i - p_j$, that is, those where $i = k$ or $j = k+1$; and those where $p_i - p_j$ was not restricted, because j was greater than $c_p + 1$ (case 4 below). There are four such cases.

Case 1. If $(i, j) = (k, k+1)$, then $q_i - q_j \geq 0 = j - i - 1$.

Case 2. If $i = k$ and $j > k+1$, apply (3) to $(i+1, j)$ to obtain

$$q_i - q_j \geq q_{i+1} - q_j = p_{i+1} - p_j + 1 \geq j - (i+1) - 1 + 1 = j - i - 1.$$

Case 3. If $j = k+1$ and $i < k$, then applying (3) to $(i, j-1)$,

$$q_i - q_j \geq q_i - q_{j-1} = p_i - p_{j-1} + 1 \geq (j-1) - i - 1 + 1 = j - i - 1.$$

Case 4. We have $j = c_p + 2 = k + 2$, $p_{k+1} = 0$ and $p_k \geq 2$. If $i = k$ or $k+1$, then $q_i - q_j = q_i \geq 1 \geq j - i - 1$. If $i < k$, then

$$q_i - q_j = p_i - 0 \geq p_i - p_k + 2 \geq (i - k - 1) + 2 = i - j - 1.$$

This concludes the inductive step and (b′) is proved.

41. Let B_n be the set of all binary strings of length n. Given two strings $(a_i)_{i=1}^n$ and $(b_i)_{i=1}^n$, define the distance between the strings as

$$d((a_i), (b_i)) = \sum_{k=1}^{n} |a_i - b_i|.$$

Let C_n be a subset of B_n. The set C_n is called a *perfect error correcting code (PECC) of length n and tolerance m* if for each string (b_i) in B_n there is a unique string (c_i) in C_n with $d((b_i), (c_i)) \leq m$. Prove that there is no PECC of length 90 and tolerance 2.

Solution: Suppose C is a PECC of length 90 and tolerance 2. Without loss of generality, assume $(0, 0, \dots, 0) \in C$. Define the weight of a string (a_i) as $\sum a_i$. There is no string (c_i) of weight 1, 2, 3, or 4 in C, or else there would exist some string (b_i) within 2 of both $(0, \dots, 0)$ and (c_i). Let n_k be the number of strings of weight k in C.

There are $\binom{90}{3}$ strings of weight 3, and each must be within distance 2 of exactly one string in C of weight 5. Each string of weight 5 is within 2 of $\binom{5}{3}$ strings of weight 3. Therefore,

$$\binom{90}{3} = \binom{5}{3} n_5, \text{ so } n_5 = 11748.$$

There are $\binom{90}{4}$ strings of weight 4, and each must be within distance 2 of exactly one string in C of weight 5 or 6. Each string of weight 5 is 1 away from exactly $\binom{5}{4}$ strings of weight 4 and 2 away from no strings of weight 4. Each string of weight 6 is 2 away from exactly $\binom{6}{4}$ strings of weight 4. Hence,

$$\binom{90}{4} = \binom{5}{4} n_5 + \binom{6}{4} n_6, \text{ so } n_6 = 116430.$$

Each of the $\binom{90}{5}$ strings of weight 5 is within 2 of exactly one string in C of weight 5, 6, or 7. A string in C of weight 5 is within 2 of itself and $85\binom{5}{4}$ other strings of weight 5. A string of weight 6 is 1 away from exactly $\binom{6}{5}$ strings of weight 5 and 2 away from no strings of weight 5. A string of weight 7 is within 2 of $\binom{7}{5}$ strings of weight 5. Thus,

$$\binom{90}{5} = \left(1 + 85\binom{5}{4}\right) n_5 + \binom{6}{5} n_6 + \binom{7}{5} n_7, \text{ so } n_7 = 1806954\tfrac{2}{7},$$

which is impossible. Therefore, there is no PECC of length 90 and tolerance 2.

42. Determine if it is possible to arrange the numbers $1, 1, 2, 2, \dots, n, n$ such that there are j numbers between two j's, $1 \le j \le n$, when $n = 2000$, $n = 2001$, and $n = 2002$. (For example, for $n = 4$, 41312432 is such an arrangement.)

Solution: We say a permutation is *good* if it satisfies the above conditions. In general, there is a good permutation of the numbers $1, 1, 2, 2, \dots, n, n$ if and only if $n = 4k$ or $4k - 1$ for some positive integer k.

First we show that there is no good permutation of the numbers $1, 1, 2, 2, \dots, n, n$ if $n \equiv 1$ or 2 modulo 4. We approach indirectly by assuming that a good permutation $a_1, a_2, \ldots, a_{2n}$ exists. For each number k, let (i_k, j_k), $i_k < j_k$, denote the positions of the two occurrences of k. Then

$$\sum_{k=1}^{n} i_k + \sum_{k=1}^{n} j_k = 1 + 2 + \cdots + 2n = n(2n+1) = S_1,$$

which is odd if $n \equiv 1 \pmod 4$ and even if $n \equiv 2 \pmod 4$. On the other hand, $j_k - i_k = k + 1$, so

$$\sum_{k=1}^{n} j_k - \sum_{k=1}^{n} i_k = 2 + 3 + \cdots + (n+1) = \frac{n(n+3)}{2} = S_2,$$

which is even if $n \equiv 1 \pmod 4$ and odd if $n \equiv 2 \pmod 4$. We obtain $2\sum_{k=1}^{n} j_k = S_1 + S_2$, an even number equal to an odd number, which is impossible.

(We can also use the following approach: Each pair of even numbers $2k$ and $2k$ $(2 \le 2k \le n)$ takes one odd position and one even position, and each pair of odd numbers $2k-1$ and $2k-1$ $(1 \le 2k-1 \le n)$ takes either two odd positions or two even positions. The odd numbers will therefore take an even number of even positions. Say that number is $2m$. Since there are $\lfloor n/2 \rfloor$ pairs of even numbers $2k$ and $2k$, even numbers will take $\lfloor n/2 \rfloor$ even positions. Hence $\lfloor n/2 \rfloor + 2m = n$, implying that $\lfloor n/2 \rfloor \equiv n \pmod 2$, which is not true for $n \equiv 1$ or 2 modulo 4.)

Now we show that there are good permutations for $n \equiv 0$ or 3 modulo 4. If $n = 3$, we have $(2, 3, 1, 2, 1, 3)$; if $n = 4$, we have $(2, 3, 4, 2, 1, 3, 1, 4)$; if $n = 4m - 1$ with $m \ge 2$, we have

$$\begin{aligned}
&[(4m-4, 4m-6, \ldots, 2m); 4m-2; (2m-3, 2m-5, \ldots, 1);\\
&4m-1; (1, 3, \ldots, 2m-3); (2m, 2m+2, \ldots, 4m-4);\\
&2m-1; (4m-3, 4m-5, \ldots, 2m+1); 4m-2;\\
&(2m-2, 2m-4, \ldots, 2); 2m-1; 4m-1;\\
&(2, 4, \ldots, 2m-2); (2m+1, 2m+3, \ldots, 4m-3)];
\end{aligned}$$

if $n = 4m$ with $m \ge 2$, we have

$$\begin{aligned}
&[(4m-2, 4m-4, \ldots, 2m); 4m-1; (2m-3, 2m-5, \ldots, 1);\\
&4m; (1, 3, \ldots, 2m-3); (2m, 2m+2, \ldots, 4m-2);\\
&2m-1; (4m-3, 4m-5, \ldots, 2m+1); 4m-1;\\
&(2m-2, 2m-4, \ldots, 2); 2m-1; 4m;\\
&(2, 4, \ldots, 2m-2); (2m+1, 2m+3, \ldots, 4m-3)].
\end{aligned}$$

Note: This problem is called the *Langford* problem. It is closely related to the following problems:

- *Langford* Determine if it is possible to partition the set

$$\{1, 2, \ldots, 2k\}$$

into k pairs of numbers $(a_1, b_1), (a_2, b_2), \ldots, (a_k, b_k)$ such that $b_i - a_i = i + 1$ for $1 \le i \le k$.

- *Skolem* Determine if it is possible to partition the set

$$\{1, 2, \ldots, 2k\}$$

into k pairs of numbers $(a_1, b_1), (a_2, b_2), \ldots, (a_k, b_k)$ such that $b_i - a_i = i$ for $1 \le i \le k$.

- *Skolem* Determine if it is possible to partition the set

$$\{2, 3, \ldots, 2k\}$$

into a singleton and $k-1$ pairs of numbers $(a_1, b_1), (a_2, b_2), \ldots, (a_k, b_k)$ such that $b_i - a_i = i$ for $1 \le i \le k$.

43. [IMO Shortlist 1996] Let k, m, n be integers such that $1 < n \le m - 1 \le k$. Determine the maximum size of a subset S of the set $\{1, 2, \ldots, k\}$ such that no n distinct elements of S add up to m.

Solution: If $m < n(n+1)/2$, then the problem is trivial: the set $\{1, 2, \ldots, k\}$ itself satisfies the stated property, so the required maximum cardinality is k. In the analysis conducted below, we assume that $m \ge n(n+1)/2$.

It is not difficult to find a lower bound on the required maximum. Let r be the largest integer such that

$$r + (r+1) + \cdots + (r + n - 1) \le m,$$

or $nr + n(n-1)/2 \le m$. It is clear that no n elements from $\{r+1, r+2, \ldots, k\}$ can add up to m. Solving the inequality yields

$$r = \left\lfloor \frac{m}{n} - \frac{n-1}{2} \right\rfloor,$$

implying that a lower bound for the required maximum is the quantity

$$k - \left\lfloor \frac{m}{n} - \frac{n-1}{2} \right\rfloor.$$

We now show that this lower bound is in fact the answer we need. To do this, it suffices to show that if S is a subset of $\{1, 2, \ldots, k\}$ and no n elements of S add up to m, then

$$|S| \leq k - \left\lfloor \frac{m}{n} - \frac{n-1}{2} \right\rfloor. \tag{$*$}$$

We induct on n. Note that $2 \leq n \leq m-1$. For the base case $n = 2$, no two elements of S can add up to m, so for each i with $1 \leq 2i \leq m-1$, at most one out of the pair of numbers $(i, m-i)$ can be in S. It follows that

$$|S| \leq k - \left\lfloor \frac{m-1}{2} \right\rfloor,$$

in agreement with $(*)$.

Let $n > 2$. We assume the result for $n-1$ and prove it for n. Let S be a subset of $\{1, 2, \ldots, k\}$ having the stated property, and let x be the least element of S. If we have $nx + n(n-1)/2 > m$, then

$$S \subseteq \{r+1, r+2, \ldots, k\},$$

with r defined as above, and there is nothing to prove. We may therefore assume that $nx + n(n-1)/2 \leq m$.

The desired property implies that $S_1 = S - \{x\}$ is a subset of $\{x+1, x+2, \ldots, k\}$ such that no $n-1$ elements of S_1 add up to $m-x$. Let $S_2 = \{s - x \mid x \in S_1\}$. Then S_2 is a subset of $\{1, 2, \ldots, k-x\}$ such that no $n-1$ distinct elements of S_2 add up to $m - nx$. To invoke the induction hypothesis, we need to show that

$$n - 1 \leq m - nx - 1 \leq k - x.$$

The first inequality is equivalent to $m - nx \geq n$, and this holds as we have already assumed that $m - nx \geq n(n-1)/2$. The second inequality is trivial as $m - 1 \leq k$. By the induction hypothesis, we obtain

$$\begin{aligned} |S| &\leq 1 + k - x - \left\lfloor \frac{m-nx}{n-1} - \frac{n-2}{2} \right\rfloor \\ &= k - \left\lfloor \frac{m-x}{n-1} - \frac{n}{2} \right\rfloor \\ &= k - \left\lfloor \frac{mn-nx}{n(n-1)} - \frac{1}{2} - \frac{n-1}{2} \right\rfloor. \end{aligned}$$

The inequality $nx + n(n-1)/2 \leq m$ implies that $mn - nx \geq (n-1)m + n(n-1)/2$. It follows that

$$\begin{aligned} |S| &\leq k - \left\lfloor \frac{(n-1)m}{n(n-1)} + \frac{1}{2} - \frac{1}{2} - \frac{n-1}{2} \right\rfloor \\ &= k - \left\lfloor \frac{m}{n} - \frac{n-1}{2} \right\rfloor, \end{aligned}$$

which is $(*)$. This completes the induction.

44. [USAMO 1998 submission, Kiran Kedlaya] A nondecreasing sequence s_0, $s_1, \ldots$ of nonnegative integers is said to be *superadditive* if $s_{i+j} \geq s_i + s_j$ for all nonnegative integers i, j. Suppose $\{s_n\}$ and $\{t_n\}$ are two superadditive sequences, and let $\{u_n\}$ be a nondecreasing sequence with the property that each integer appears in $\{u_n\}$ as many times as in $\{s_n\}$ and $\{t_n\}$ combined. Show that $\{u_n\}$ is also superadditive.

Solution: Given a sequence $\{a_n\}$, define the dual sequence $\{S_n\}$ so that S_n equals the smallest integer k such that $s_k \geq n$. The key observation is that $\{s_n\}$ is superadditive if and only if $\{S_n\}$ is *subadditive*, that is, $S_{i+j} \leq S_i + S_j$. Indeed, assume that $\{s_n\}$ is superadditive. Then

$$s_{S_i+S_j} \geq s_{S_i} + s_{S_j} \geq i + j,$$

and so by the definition of S, $S_{i+j} \leq S_i + S_j$; the reverse implication is proved similarly.

Now simply note that if $\{S_n\}$ and $\{T_n\}$ are dual sequences of $\{s_n\}$ and $\{t_n\}$, respectively, then the dual sequence of $\{u_n\}$ is $\{S_n + T_n\}$. Since $\{S_n\}$ and $\{T_n\}$ are subadditive, clearly $\{S_n + T_n\}$ is as well, and so $\{u_n\}$ is superadditive, as desired.

45. [IMO Shortlist 1999] The numbers from 1 to n^2, $n \geq 2$, are randomly arranged in the cells of an $n \times n$ unit square grid. For any pair of numbers situated on the same row or on the same column, the ratio of the greater number to the smaller one is calculated. The *characteristic* of the arrangement is the smallest of these $n^2(n-1)$ fractions. Determine the largest possible value of the characteristic.

Solution: We first show that for any arrangement A its characteristic $C(A)$ is less than or equal to $(n+1)/n$. If two numbers in the set $G = \{n^2 - n + 1, n^2 - n + 2, \ldots, n^2\}$ lie on the same row or column, then

$$C(A) \leq \frac{n^2}{n^2 - n + 1} < \frac{n+1}{n}.$$

If the numbers in G are in different rows and columns, then two of them are on the same row or column as the number $n^2 - n$, so we have

$$C(A) \le \frac{n^2-1}{n^2-n} = \frac{n+1}{n}.$$

Now we show that the arrangement

$$a_{ij} = \begin{cases} i + n(j-i-1) & \text{if } i < j, \\ i + n(n-i+j-1) & \text{if } i \ge j, \end{cases}$$

that is,

$1+(n-1)n$	1	$\dots$	$1+(n-2)n$
$2+(n-2)n$	$2+(n-1)n$	$\dots$	$2+(n-3)n$
$3+(n-3)n$	$3+(n-2)n$	$\dots$	$3+(n-4)n$
$\vdots$	$\vdots$	$\vdots$	$\vdots$
$(n-2)+2n$	$(n-2)+3n$	$\dots$	$(n-2)+n$
$(n-1)+n$	$(n-1)+2n$	$\dots$	$n-1$
n	$n+n$	$\dots$	$n+(n-1)n$

has characteristic $(n+1)/n$. Indeed,

- The difference between any two numbers lying on the same row is a multiple of n; therefore,

$$\frac{a_{ik}}{a_{ij}} = \frac{a_{ik}}{a_{ik}-hn} \ge \frac{a_{ik}}{a_{ik}-n} \ge \frac{n^2}{n^2-n} > \frac{n+1}{n};$$

- On the first column we have the arithmetic progression

$$n \le (n-1)+n \le (n-2)+2n \le \\ \cdots \le 2+(n-2)n \le 1+(n-1)n,$$

implying

$$\frac{a_{i1}}{a_{k1}} \ge \frac{1+(n-1)n}{2+(n-2)n} = \frac{n^2-n+1}{n^2-2n+2} \ge \frac{n+1}{n},$$

with equality if $n = 2$;

- The jþ column, $2 \leq j \leq n$, contains the two arithmetic progressions:

$$j-1, (j-2)+n, (j-3)+2n, \ldots, 1+(j-2)n;$$
$$n+(j-1)n, (n-1)+jn, \ldots, j+(n-1)n,$$

implying

$$\frac{a_{ij}}{a_{kj}} \geq \frac{j+(n-1)n}{(j+1)+(n-2)n} \geq \frac{n+1}{n},$$

with equality for $j = n-1$.

46. [China 1999, Hongbin Yu] For a set S, let $|S|$ denote the number of elements in S. Let A be a set with $|A| = n$, and let $A_1, A_2, \ldots, A_n$ be subsets of A with $|A_i| \geq 2$, $1 \leq i \leq n$. Suppose that for each 2-element subset A' of A, there is a unique i such that $A' \subseteq A_i$. Prove that $A_i \cap A_j \neq \emptyset$ for all $1 \leq i < j \leq n$.

Solution: By the given conditions, we have

$$\sum_{i=1}^{n} \binom{|A_i|}{2} = \binom{n}{2}. \tag{4}$$

Let $A = \{x_1, x_2, \ldots, x_n\}$, and let d_i denote the number of subsets A_j, $1 \leq j \leq n$, such that $x_i \in A_j$. Hence

$$\sum_{i=1}^{n} d_i = \sum_{i=1}^{n} |A_i|. \tag{5}$$

On the other hand,

$$\sum_{i=1}^{n} \binom{d_i}{2} = \sum_{1 \leq i < j \leq n} |A_i \cap A_j|.$$

By the conditions of the problem, $|A_i \cap A_j| \leq 1$. It suffices to prove that $|A_i \cap A_j| = 1$, or

$$\sum_{i=1}^{n} \binom{d_i}{2} = \binom{n}{2}.$$

By (4) and (5) and by the definition of the binomial coefficient $\binom{x}{2} = \frac{x^2-x}{2}$, it suffices to prove

$$\sum_{i=1}^{n} d_i^2 = \sum_{i=1}^{n} |A_i|^2. \tag{6}$$

For each x_i, we consider the sets A_j such that $x_i \notin A_j$. Let $A_j = \{y_1, y_2, \ldots, y_s\}$ be one such set. Since each of the 2-element sets $\{x_i, y_1\}$, $\{x_i, y_2\}$, $\ldots$, $\{x_i, y_s\}$ is a subset of a distinct set A_k (as y_i and y_j cannot both be in another set again), $d_i \geq |A_j|$. It follows that

$$\frac{d_i}{n - d_i} \geq \frac{|A_j|}{n - |A_j|}.$$

Summing up all the above inequalities yields

$$\begin{aligned} \sum_{i=1}^{n} d_i &= \sum_{i=1}^{n} \sum_{j \mid x_i \notin A_j} \frac{d_i}{n - d_i} \geq \sum_{i=1}^{n} \sum_{j \mid x_i \notin A_j} \frac{|A_j|}{n - |A_j|} \\ &= \sum_{j=1}^{n} \sum_{i \mid x_i \notin A_j} \frac{|A_j|}{n - |A_j|} \\ &= \sum_{j=1}^{n} |A_j|. \end{aligned}$$

By (5), all the equalities hold in the above inequalities. Hence $d_i = |A_j|$. It follows that

$$\begin{aligned} \sum_{i=1}^{n} (n - d_i) d_i &= \sum_{i=1}^{n} \sum_{j \mid x_i \notin A_j} d_i = \sum_{i=1}^{n} \sum_{j \mid x_i \notin A_j} |A_j| \\ &= \sum_{j=1}^{n} \sum_{i \mid x_i \notin A_j} |A_j| \\ &= \sum_{j=1}^{n} (n - |A_j|)|A_j| \end{aligned}$$

implying (6), as desired.

47. [Iran 1999] Suppose that $r_1, \ldots, r_n$ are real numbers. Prove that there exists a set $S \subseteq \{1, 2, \ldots, n\}$ such that

$$1 \leq |S \cap \{i, i+1, i+2\}| \leq 2,$$

for $1 \leq i \leq n - 2$, and

$$\left| \sum_{i \in S} r_i \right| \geq \frac{1}{6} \sum_{i=1}^{n} |r_i|.$$

Solution: Let $s = \sum_{i=1}^{n} |r_i|$ and for $i = 0, 1, 2$, define

$$s_i = \sum_{r_j \geq 0, j \equiv i \pmod 3} r_j \quad \text{and} \quad t_i = \sum_{r_j < 0, j \equiv i \pmod 3} r_j.$$

Then we have $s = s_1 + s_2 + s_3 - t_1 - t_2 - t_3$, or

$$\begin{aligned} 2s &= (s_1 + s_2) + (s_2 + s_3) + (s_3 + s_1) \\ &\quad -(t_1 + t_2) - (t_2 + t_3) - (t_3 + t_1). \end{aligned}$$

Therefore there are $i_1 \neq i_2$ such that either $s_{i_1} + s_{i_2} \geq \frac{s}{3}$ or $t_{i_1} + t_{i_2} \leq -\frac{s}{3}$ or both. Without loss of generality, we assume that $s_{i_1} + s_{i_2} \geq \frac{s}{3}$ and $|s_{i_1} + s_{i_2}| \geq |t_{i_1} + t_{i_2}|$. Thus $s_{i_1} + s_{i_2} + t_{i_1} + t_{i_2} \geq 0$. We have

$$(s_{i_1} + s_{i_2} + t_{i_1}) + (s_{i_1} + s_{i_2} + t_{i_2}) \geq s_{i_1} + s_{i_2} \geq \frac{s}{3}.$$

Therefore at least one of $s_{i_1} + s_{i_2} + t_{i_1}$ and $s_{i_1} + s_{i_2} + t_{i_2}$ is greater than or equal to $\frac{s}{6}$ and we are done.

Note: By setting $r_1 = r_2 = r_3 = 1$ and $r_4 = r_5 = r_6 = -1$, it is easy to prove that $\frac{1}{6}$ is the best value for the bound.

48. [USAMO 1999 submission, Kiran Kedlaya] Let n, k, m be positive integers with $n > 2k$. Let S be a nonempty set of k-element subsets of $\{1, \ldots, n\}$ with the property that every $(k+1)$-element subset of $\{1, \ldots, n\}$ contains exactly m elements of S. Prove that S must contain every k-element subset of $\{1, \ldots, n\}$.

Solution: We first count pairs of (U, V), where $U \in S$ and V is a $(k+1)$-element subset of $\{1, 2, \ldots, n\}$ containing U, in two different ways. By assumption, if we sort the pairs by V, we find there are $m\binom{n}{k+1}$ of them. On the other hand, if we sort by U, we find there are $(n-k)|S|$ pairs. We conclude

$$|S| = \frac{m}{n-k}\binom{n}{k+1} = \frac{m}{k+1}\binom{n}{k}.$$

We next count triples (U, V, W), where U is a $(k+1)$-element subset of $\{1, 2, \ldots, n\}$ and V and W are distinct elements of S contained in U. If we sort the triples by U, we see that by assumption the number of triples is

$$m(m-1)\binom{n}{k+1} = \frac{m(m-1)(n-k)(n-k+1)}{k(k+1)}\binom{n}{k-1}.$$

On the other hand, we can also sort the triples by $V \cap W$, which is always a set of $k-1$ elements. For each $(k-1)$-element subset J of $\{1, 2, \ldots, n\}$, let s_J be the number of elements of S containing it. Each J is $V \cap W$ for exactly $s_J(s_J - 1)$ triples, so we conclude

$$\frac{m(m-1)(n-k)(n-k+1)}{k(k+1)}\binom{n}{k-1} = \sum_J s_J(s_J - 1).$$

The function $f(x) = x^2$ is convex, and

$$\sum_J s_J = k|S| = \frac{mk}{k+1}\binom{n}{k} = \frac{m(n-k+1)}{k+1}\binom{n}{k-1}.$$

Hence

$$\begin{aligned} &\frac{m(m-1)(n-k)(n-k+1)}{k(k+1)}\binom{n}{k-1} \\ \geq\ &\frac{m^2(n-k+1)^2}{(k+1)^2}\binom{n}{k-1} - \frac{m(n-k+1)}{k+1}\binom{n}{k-1}, \end{aligned}$$

or

$$\frac{(m-1)(n-k)}{k} \geq \frac{m(n-k+1)}{k+1} - 1.$$

Hence

$$m(n-2k) \geq (k+1)(n-2k).$$

Since we assume that $n > 2k$, we conclude that $m \geq k+1$, which implies that S must contain every k-element subset of $\{1, 2, \ldots, n\}$.

Note: The result is a special case of a theorem of Liningston and Kantor, but with a different proof. The condition $n > 2k$ is definitely necessary, or else balanced block designs would not exist. Complementing the sets of such a design gives a counterexample.

49. [China 1999, Zhenhua Qu] A set T is called *even* if it has an even number of elements. Let n be a positive even integer, and let $S_1, S_2, \ldots, S_n$ be even subsets of the set $S = \{1, 2, \ldots, n\}$. Prove that there exist i and j, $1 \leq i < j \leq n$, such that $S_i \cap S_j$ is even.

First Solution: For a pair of sets A and B, we introduce their *symmetric difference*

$$A \Delta B = (A - B) \cup (B - A),$$

that is, $A\Delta B = (A \cup B) - (A \cap B)$. For each subset $A \subseteq T$, we define its *index function* $\psi_A : T \to \{0, 1\}$ as

$$\psi_A(x) = \begin{cases} 1 & \text{if } x \in A \\ 0 & \text{if } x \notin A. \end{cases}$$

We have the following properties of the symmetric difference operation:

(a) $A\Delta A = \emptyset$ and $A\Delta\emptyset = A$;

(b) $A\Delta B = B\Delta A$;

(c) $(A\Delta B)\Delta C = A\Delta(B\Delta C)$;

(d) if both A and B are even sets, $A\Delta B$ is also even;

(e) $a \in A_1\Delta A_2\Delta\cdots\Delta A_r$ (r a positive integer) if and only if x belongs to an odd number of sets $A_1, A_2, \ldots, A_r$.

The proofs of properties (a), (b), (c), (d) are rather straightforward. We prove property (e) by induction on r. The base cases $r = 1$ and $r = 2$ are trivial. Assume the property is true for r, $r \geq 2$, for the subsets $A_1, A_2, \ldots, A_r$. We consider

$$X = A_1\Delta A_2\Delta\cdots A_r\Delta A_{r+1} = X_1\Delta A_{r+1},$$

where $X_1 = A_1\Delta A_2\Delta\cdots A_r$. Then $x \in X$ if and only if x belongs to either $X_1 - A_{r+1}$ or $A_{r+1} - X$. If $x \in X_1 - A_{r+1}$, $x \notin A_{r+1}$ and $x \in X$; that is, x belongs to an odd number of sets $A_1, A_2, \ldots, A_r, A_{r+1}$ by the induction hypothesis. If $x \in A_{r+1} - X$, then $x \in A_{r+1}$ and x belongs to an even number of sets $A_1, A_2, \ldots, A_r$, so x belongs to an odd number of sets $A_1, A_2, \ldots, A_r, A_{r+1}$. Our induction is complete.

Now we are ready to prove our main result. Let $n = 2m$, and let

$$\mathcal{S} = \{S_1, S_2, \ldots, S_{2m}\}.$$

We have the following lemma.

Lemma. *There exist an even number of sets S_i with their symmetric difference equal to either $\emptyset$ or S.*

Proof: We consider all possible symmetric differences

$$S_{i_1}\Delta S_{i_2}\Delta\cdots\Delta S_{i_{2j}}$$

where $j \geq 1$. If some of these difference are equal to either $\emptyset$ or S, we are done; if not, note that there are $2^{2m-1} - 1$ such differences and each of these is an even subset of S (by property (d)). Also note that there are $2^{2m-1} - 2$ distinct even subset of S not including $\emptyset$ and S. By the Pigeonhole

Principle, two of these differences are the same. Without loss of generality (by properties (b) and (c)), we may assume

$$\begin{aligned} T_1 &= S_1 \Delta S_2 \Delta \cdots \Delta S_k \Delta S_{k+1} \Delta \cdots \Delta S_{2i} \\ &= S_{k+1} \Delta \cdots \Delta S_{2i} \Delta S_{2i+1} \Delta S_{2j-k} = T_2 \end{aligned}$$

By properties (a) and (b), we obtain

$$\begin{aligned} \emptyset &= T_1 \Delta T_2 \\ &= S_1 \Delta S_2 \Delta \cdots \Delta S_k \Delta S_{2i+1} \Delta \cdots \Delta S_{2j-k}. \end{aligned}$$

Hence $\emptyset$ is the symmetric difference of $k + (2j - k - 2i) = 2(j - i)$ sets. ■

By the lemma, we may assume that there is an even number $2i$ such that

$$S_1 \Delta S_2 \cdots \Delta S_{2i} = \emptyset \quad \text{or} \quad S.$$

We consider these two cases separately.

- *Case 1.* We assume that

$$S_1 \Delta S_2 \cdots \Delta S_{2i} = \emptyset.$$

By property (e), each element s of S belongs to an even number of sets $S_1, S_2, \ldots, S_{2i}$. We calculate

$$r_1 = \sum_{k=2}^{2i} |S_1 \cap S_k| = \sum_{k=2}^{2i} \left(\sum_{s_1 \in S_1} \psi_{S_k}(s_1) \right).$$

For $s_1 \in S_1$, s_1 belongs to an odd number of sets $S_2, \ldots, S_{2i}$, i.e.,

$$\sum_{k=2}^{2i} \psi_{S_k}(s_1) \equiv 1 \pmod 2.$$

Therefore,

$$r_1 = \sum_{k=2}^{2i} \left(\sum_{s_1 \in S_1} \psi_{S_k}(s_1) \right) = \sum_{s_1 \in S_1} \left(\sum_{k=2}^{2i} \psi_{S_k}(s_1) \right).$$

Since S_1 is an even set, r_1 is even. Therefore, at least one of the $2i - 1$ numbers $|S_1 \cap S_k|$, $2 \le k \le 2i$, must be even, as desired.

- *Case 2.* We assume that

$$S_1 \Delta S_2 \cdots \Delta S_{2i} = S.$$

By property (e), each element s of S belongs to an odd number of sets $S_1, S_2, \ldots, S_{2i}$. We calculate

$$r_1 = \sum_{k=2}^{2i} |S_1 \cap S_k| = \sum_{k=2}^{2i} \left(\sum_{s_1 \in S_1} \psi_{S_k}(s_1) \right).$$

For $s_1 \in S_1$, s_1 belongs to an odd number of sets $S_2, \ldots, S_{2i}$, i.e.,

$$\sum_{k=2}^{2i} \psi_{S_k}(s_1) \equiv 0 \pmod{2}.$$

Therefore,

$$r_1 = \sum_{k=2}^{2i} \left(\sum_{s_1 \in S_1} \psi_{S_k}(s_1) \right) = \sum_{s_1 \in S_1} \left(\sum_{k=2}^{2i} \psi_{S_k}(s_1) \right).$$

Since r_1 is the sum of even numbers, r_1 is even. Therefore, at least one of the $2i - 1$ numbers $|S_1 \cap S_k|$, $2 \le k \le 2i$, must be even, as desired.

Second Solution: (By Tiankai Liu) Assign to each S_i an n-dimensional vector a_i, where the jþ coordinate of a_i is 1 if S_i contains j and 0 if not. Then the fact that S_i is even translates into $a_i \cdot a_i \equiv 0 \pmod{2}$.

For the sake of contradiction, assume $a_i \cdot a_j \equiv 1 \pmod{2}$ for all $i \ne j$, that is, S_i and S_j have odd intersection for all $i \ne j$. Suppose there exists a nonempty subset X of $\{1, 2, 3, \ldots, n\}$ for which

$$\sum_{x \in X} a_x \equiv (0, 0, 0, \ldots, 0) \pmod{2}.$$

Then, for any $i \in X$, $a_i \cdot \sum_{x \in X} a_x \equiv 0 \pmod{2}$. On the other hand,

$$\begin{aligned} a_i \cdot \sum_{x \in X} a_x &= a_i \cdot a_i + a_i \cdot \sum_{x \in (X-i)} a_x \\ &\equiv a_i \cdot \sum_{x \in (X-i)} \\ &\equiv |X| - 1 \pmod{2}. \end{aligned}$$

It follows that that $|X|$ is odd. So X is a proper subset of $\{1, 2, \ldots, n\}$, since n is even. But then, for any $j \notin X$, $a_j \cdot \sum_{x \in X} a_x \equiv 0 \pmod 2$ implies that $|X|$ is even, a contradiction. Thus X does not exist.

Thus, for each of the 2^n subsets Y of $\{1, 2, 3, \ldots, n\}$, $\sum_{y \in Y} a_y$ is different modulo 2. On the other hand, for each such sum, the sum of all the coordinates of the vector must be even, as it was formed by summing vectors corresponding to even sets. Hence the first $n - 1$ coordinates of the vector determine the parity of the last coordinate. It follows that there are only 2^{n-1} possibilities, modulo 2, for $\sum_{y \in Y} a_y$, a contradiction.

Thus, there exist i and j such that $a_i \cdot a_j$ is even, and correspondingly the intersection of S_i and S_j is even.

50. Let $A_1, A_2, \ldots, B_1, B_2, \ldots$ be sets such that $A_1 = \emptyset$, $B_1 = \{0\}$,

$$A_{n+1} = \{x + 1 \mid x \in B_n\}, \quad B_{n+1} = A_n \cup B_n - A_n \cap B_n,$$

for all positive integers n. Determine all the positive integers n such that $B_n = \{0\}$.

First Solution: We show that $B_n = 0$ if and only if n is a power of 2.

First we introduce some notation. For a set S of integers, let $2S$ denote the set $\{2x \mid x \in S\}$, and let $S + k$ denote the set $\{x + k \mid x \in S\}$ for any integer k.

We now observe that $0 \notin A_n$ for all $n \geq 1$. This is true for A_1 by definition, and for the rest because A_{n+1} is formed by adding 1 to the elements of B_n, all of which are nonnegative. It follows by an easy induction that $0 \in B_n$ for all $n \geq 1$.

Next we prove by induction that the following four statements are true for all $n \geq 2$:

(a) $A_{2n-1} = 2A_n - 1$;
(b) $B_{2n-1} = A_{2n-1} \cup B_{2n}$;
(c) $B_{2n} = 2B_n$;
(d) $1 \in B_{2n-1}$.

The base case can be verified by computing $A_2 = \{1\}$, $B_2 = \{0\}$, $A_3 = \{1\}$, $B_3 = \{0, 1\}$, $A_4 = \{1, 2\}$, and $B_4 = \{0\}$. For the induction step, assume that the statement holds for $n - 1$; that is, assume that $A_{2n-3} = 2A_{n-1} - 1$, $B_{2n-3} = A_{2n-3} \cup B_{2n-2}$, $B_{2n-2} = 2B_{n-1}$, and $1 \in B_{2n-3}$. Then statement (a) immediately follows because

$$\begin{aligned} A_{2n-1} &= B_{2n-2} + 1 = 2B_{n-1} + 1 \\ &= 2(A_n - 1) + 1 = 2A_n - 1, \end{aligned}$$

where we used the definition in steps 1 and 3 and the induction hypothesis in step 2.

Moving on to (b) and (c), we first need to obtain some information about A_{2n-2}:

$$\begin{aligned} A_{2n-2} &= B_{2n-3}+1=(A_{2n-3}\cup B_{2n-2})+1 \\ &= ((2A_{n-1}-1)\cup 2B_{n-1})+1=2A_{n-1}\cup(2B_{n-1}+1). \end{aligned}$$

Using this, we can compute B_{2n-1}. We know by assumption that $B_{2n-2}=2B_{n-1}$, and we have by definition that $B_{2n-1}=A_{2n-2}\cup B_{2n-2}-A_{2n-2}\cap B_{2n-2}$ and $B_n=A_{n-1}\cup B_{n-1}-A_{n-1}\cap B_{n-1}$. Therefore,

$$\begin{aligned} B_{2n-1} &= 2A_{n-1}\cup(2B_{n-1}+1)\cup 2B_{n-1} \\ &\quad -(2A_{n-1}\cup(2B_{n-1}+1))\cap 2B_{n-1} \\ &= (2B_{n-1}+1)\cup 2A_{n-1}\cup 2B_{n-1}-2A_{n-1}\cap 2B_{n-1} \\ &= (2B_{n-1}+1)\cup(2A_{n-1}\cup 2B_{n-1}-2A_{n-1}\cap 2B_{n-1}) \\ &= (2B_{n-1}+1)\cup 2B_n, \end{aligned}$$

where we could remove the set $2B_{n-1}+1$ in the second step because it contains only odd elements, while $2B_{n-1}$ contains only even elements. We saw above that $A_{2n-1}=2B_{n-1}+1$, so we have $B_{2n-1}=A_{2n-1}\cup 2B_n$.

Now we are finally ready to prove (c), and from there (b). We know by definition that B_{2n} consists of the numbers that are either in A_{2n-1} or B_{2n-1} but not both. We just proved that $B_{2n-1}=A_{2n-1}\cup 2B_n$, so B_{2n} cannot contain any elements in A_{2n-1}. On the other hand, we also showed that $A_{2n-1}=2A_n-1$, so all the elements of A_{2n-1} are odd and they cannot coincide with the elements of $2B_n$. Therefore, B_{2n} is exactly $2B_n$, proving (c). Statement (b) now follows immediately, simply by plugging in $B_{2n}=2B_n$ in the relation $B_{2n-1}=A_{2n-1}\cup 2B_n$.

We can now prove (d) by looking back at the aftermath. We showed above that $B_{2n-1}=(2B_{n-1}+1)\cup 2B_n$. At the very beginning we observed that 0 is an element of each of the B's. It follows that 1 is an element of $2B_{n-1}+1$, and hence of B_{2n-1}. This proves (d).

At this point it is easy to conclude that the integers n such that $B_n=\{0\}$ are exactly the powers of 2. To show that the powers of 2 have this property, only a trivial induction using the relation $B_{2n}=2B_n$ is required. To show that no other numbers do, observe that all B's with odd indices contain 1 by (d), and another trivial induction using the same relation suffices.

Second Solution: Let g_n be generating functions defined by $g_1(x)=g_2(x)=1$, and $g_{n+1}(x)=g_n(x)+xg_{n-1}(x)$. We claim that for each i,

$i \in B_n$ if and only if the coefficient of x^i in $g_n(x)$ is odd. To prove this, first define a corresponding sequence of generating functions f_n for A_n by $f_1(x) = 0$, $f_{n+1}(x) = xg_n(x)$. We will show also that the f's represent the A's by the same scheme. We do so by induction. Clearly, the representations are correct for $n = 1$. Now we assume that they are correct for some $n \geq 1$, and prove that they work for $n + 1$. The definition $f_{n+1}(x) = xg_n(x)$ is exactly the translation of the given definition $A_{n+1} = \{x + 1 \mid x \in B_n\}$. As for B_{n+1}, to keep with the given $B_{n+1} = A_n \cup B_n - A_n \cap B_n$, we can choose $g_{n+1}(x) = g_n(x) + f_n(x)$. For $n \geq 2$, we may then plug in $f_n(x) = xg_{n-1}(x)$ to obtain $g_{n+1}(x) = g_n(x) + xg_{n-1}(x)$, as wanted.

We claim that $g_n(x) = \sum_{k=0}^{\infty} \binom{n-1-k}{k} x^k$ satisfies the above recursion. Indeed, by this definition,

$$\begin{aligned} g_n(x) + xg_{n-1}(x) &= \sum_{k=0}^{\infty} \binom{n-1-k}{k} x^k + x \sum_{k=0}^{\infty} \binom{n-2-k}{k} x^k \\ &= \sum_{k=0}^{\infty} \binom{n-1-k}{k} x^k + \sum_{k=1}^{\infty} \binom{n-1-k}{k-1} x^k \\ &= \sum_{k=0}^{\infty} \binom{n-k}{k} x^k = g_{n+1}(x). \end{aligned}$$

It remains to show that all but the constant coefficient of $g_n(x)$ are even if and only if n is a power of 2. First we prove the "if" portion. Let $n = 2^m$. We must show that for all $0 < k < 2^{m-1}$,

$$\binom{2^m - 1 - k}{k} = \frac{(2^m - 1 - k)(2^m - 2 - k) \cdots (2^m - 2k)}{k \cdot (k-1) \cdots 1}$$

is even. Both the numerator and the denominator contain a product of k consecutive integers. Observe that since the product in the denominator starts at 1, for any integer a, the numerator contains at least as many multiples of a as the denominator. In particular, this holds for powers of 2. In fact, we can even show a bit more: if 2^l is the highest power of 2 dividing k, then the numerator contains strictly more multiples of 2^{l+1} than the denominator. The denominator clearly contains $\lfloor \frac{k}{2^{l+1}} \rfloor$ such multiples. The numerator, however, contains $\lceil \frac{k}{2^{l+1}} \rceil$ multiples, $2^m - 2k$ being the smallest of them. Since 2^{l+1} does not divide k by assumption, the numerator contains one more multiple of 2^{l+1} than the denominator. Now, recall that the highest power of 2 that divides a product is equal to the sum over the number of multiples of 2^j in the product, $j = 1, 2, \ldots$. Hence it follows from the above discussion that the required binomial coefficient is indeed even.

Now we prove the other direction: if n is not a power of 2, then at least one of the binomial coefficients other than the first is odd. Let $n = 2^m p$, where $p > 1$ is odd. Then consider the term

$$\binom{n-1-2^m}{2^m} = \frac{(n-1-2^m)(n-2-2^m)\cdots(n-2^{m+1})}{1\cdot 2\cdots 2^m}.$$

Since $2^m \mid n$,

$$n - i - 2^m \equiv -i \pmod{2^m}$$

for all i. Thus, for all $i < 2^m$, the factor $n - i - 2^m$ in the numerator has exactly the same number of factors of 2 as the factor i in the denominator. The same is true for the last pair of factors, $n - 2^{m+1}$ and 2^m. Neither is divisible by 2^{m+1} while both are divisible by 2^m, so the binomial coefficient $\binom{n-1-2^m}{2^m}$ is indeed odd.

51. [Iran 1999] Suppose that $S = \{1, 2, \ldots, n\}$ and that $A_1, A_2, \ldots, A_k$ are subsets of S such that for every $1 \le i_1, i_2, i_3, i_4 \le k$, we have

$$|A_{i_1} \cup A_{i_2} \cup A_{i_3} \cup A_{i_4}| \le n - 2.$$

Prove that $k \le 2^{n-2}$.

Solution: For a set T, let $|T|$ denote the numbers of elements in T. We call a set $T \subseteq S$ *2-coverable* if $T \subseteq A_i \cup A_j$ for some i and j (not necessarily distinct). By the given condition, for any subset T of S at least one of the sets T and $S - T$ is not 2-coverable. Among the subsets of S that are not 2-coverable, let A be a subset with minimum $|A|$.

Consider the family of sets $S_1 = \{A \cap A_1, A \cap A_2, \ldots, A \cap A_k\}$. ($A \cap A_i$ might equal $A \cap A_j$ for some distinct i and j, but we ignore any duplicate sets.) Because A is not 2-coverable, if $X \in S_1$, then $A - X \notin S_1$. Thus, at most half the subsets of $|A|$ are in S_1, and $|S_1| \le 2^{|A|-1}$.

On the other hand, let $B = S - A$ and consider the family of sets $S_2 = \{B \cap A_1, B \cap A_2, \ldots, B \cap A_k\}$. We claim that if $X \in S_2$, then $B - X \notin S_2$. Suppose on the contrary that both $X, B - X \in S_2$ for some $X = B \cap A_\ell$ and $B - X = B \cap A_{\ell'}$. By the minimal definition of A, there are A_i and A_j such that $A_i \cup A_j = A \setminus \{m\}$ for some i, j, and m. Then

$$|A_\ell \cup A_{\ell'} \cup A_i \cup A_j| = n - 1,$$

a contradiction. Thus our assumption is false and $|S_2| \le 2^{|B|-1} = 2^{n-|A|-1}$.

Because every set A_i is uniquely determined by its intersection with sets A and $B = S - A$, it follows that $k \le |S_1| \cdot |S_2| \le 2^{n-2}$.

Glossary

Binomial Coefficient

$$\binom{n}{k} = \frac{n!}{k!(n-k)!},$$

the coefficient of x^k in the expansion of $(x+1)^n$.

Complete Graph

Given a set V of vertices, the graph formed by joining each pair of vertices in V is the called the complete graph on V and denoted by K_V. For positive integer n, K_n denote a complete graph of n vertices.

Dirichlet's Theorem

A set S of primes is said to have *Dirichlet density* if

$$\lim_{s \to 1} \frac{\sum_{p \in S} p^{-s}}{\ln(s-1)^{-1}}$$

exists. If the limit exists we set it equal to $d(S)$ and call $d(S)$ the Dirichlet density of S.

There are infinitely many primes in any arithmetic sequence of integers for which the common difference is relatively prime to the terms. In other words, let a and m be relatively prime positive integers, then there are infinitely many primes p such that $p \equiv a \pmod{m}$. More precisely, let $S(a; m)$ denote the set of all such primes, we have $d(S(a; m)) = 1/\phi(m)$, where ϕ is Euler's function.

Euler Function

Let n be a positive integer. The Euler function $\phi(n)$ is defined to be the number of integers between 1 and n that are relatively prime to n. The following are three fundamental properties of this function:

- $\phi(nm) = \phi(n)\phi(m)$ for positive integers m and n;
- if $n = p_1^{a_1} p_2^{a_2} \cdots p_k^{a_k}$ is a prime factorization of n (with distinct primes p_i), then
$$\phi(n) = n\left(1 - \frac{1}{p_1}\right)\left(1 - \frac{1}{p_2}\right)\cdots\left(1 - \frac{1}{p_k}\right);$$
- $\displaystyle\sum_{d\|n} \phi(d) = n.$

Fermat's Little Theorem

If p is prime, then $a^p \equiv a \pmod{p}$ for all integers a.

Fibonacci Numbers

Sequence defined recursively by

$$F_1 = F_2 = 1, \quad F_{n+1} = F_n + F_{n-1}$$

for all $n \geq 2$. More explicitly,

$$F_n = \frac{1}{\sqrt{5}}\left[\left(\frac{1+\sqrt{5}}{2}\right)^n - \left(\frac{1-\sqrt{5}}{2}\right)^n\right]$$

for all $n \geq 1$.

Hamiltonain Cycles

A *walk* in a graph G is a finite sequence of vertics $v_0, v_1, \ldots, v_n$ with edges $v_0v_1, v_1v_2, \ldots, v_{n-1}v_n$. Vertices v_0 and v_n are the *end points* of the walk. A *simple walk* is a walk in which no edges is repeated. A walk is closed if the end points of the walk are the same. A closed simple work is a *cycle* if $n \geq 3$ and $v_0, v_2, \ldots, v_{n-1}$ are all different. Here n is called the *length* of the cycle. A cycle that passes through every vertex in a graph is a Hamilton cycle.

Lucas Numbers

Sequence defined recursively by

$$L_1 = 1, \quad L_2 = 3, \quad L_{n+1} = L_n + L_{n-1}$$

for all $n \geq 2$. More explicitly,

$$L_n = \left[\left(\frac{1+\sqrt{5}}{2} \right)^n + \left(\frac{1-\sqrt{5}}{2} \right)^n \right]$$

for all $n \geq 1$.

Monochromatic

Suppose the edges of a graph G are colored in k colors. We say a subgraph H of a graph G is *monochromatic* if all its edges are colored the same color.

Permutation

Let S be a set. A permutation of S is a one-to-one function $\pi : S \to S$ that maps S onto S. If $S = \{x_1, x_2, \ldots, x_n\}$ is a finite set, then we may denote a permutation π of S by $\{y_1, y_2, \ldots, y_n\}$, where $y_k = \pi(x_k)$.

Pigeonhole Principle

If n objects are distributed among $k < n$ boxes, some box contains at least two objects.

[illegible] with no edge is repeated. A walk is closed if the end points of [illegible] [illegible] through [illegible] a Hamiltonian cycle.

[illegible]

Monotonic [illegible]

[illegible] graph [illegible]

Permutation

Let S be [illegible] that maps [illegible] may denote a permutation [illegible]

[illegible]

Further Reading

1. Andreescu, T.; Kedlaya, K.; Zeitz, P., *Mathematical Contests 1995–1996: Olympiad Problems from around the World, with Solutions*, American Mathematics Competitions, 1997.

2. Andreescu, T.; Kedlaya, K., *Mathematical Contests 1996–1997: Olympiad Problems from around the World, with Solutions*, American Mathematics Competitions, 1998.

3. Andreescu, T.; Kedlaya, K., *Mathematical Contests 1997–1998: Olympiad Problems from around the World, with Solutions*, American Mathematics Competitions, 1999.

4. Andreescu, T.; Feng, Z., *USA and International Mathematical Olympiads 2000* , Mathematical Association of America, 2001.

5. Andreescu, T.; Feng, Z., *101 Problems in Algebra*, Australian Mathematics Trust, 2001.

6. Andreescu, T.; Feng, Z., *Mathematical Olympiads: Problems and Solutions from around the World, 1998–1999*, Mathematical Association of America, 2000.

7. Andreescu, T.; Feng, Z., *Mathematical Olympiads: Problems and Solutions from around the World, 1999–2000*, Mathematical Association of America, 2001.

8. Andreescu, T.; Gelca, R., *Mathematical Olympiad Challenges*, Birkhäuser, 2000.
9. Cofman, J., *What to Solve?*, Oxford Science Publications, 1990.
10. Doob, M., *The Canadian Mathematical Olympiad 1969–1993*, University of Toronto Press, 1993.
11. Engel, A., *Problem-Solving Strategies*, Problem Books in Mathematics, Springer, 1998.
12. Fomin, D.; Kirichenko, A., *Leningrad Mathematical Olympiads 1987–1991*, MathPro Press, 1994.
13. Fomin, D.; Genkin, S.; Itenberg, I., *Mathematical Circles*, American Mathematical Society, 1996.
14. Graham, R. L.; Knuth, D. E.; Patashnik, O., *Concrete Mathematics*, Addison-Wesley, 1989.
15. Greitzer, S. L., *International Mathematical Olympiads, 1959-1977*, New Mathematical Library, Vol. 27, Mathematical Association of America, 1978.
16. Grossman, I.; Magnus, W., *Groups and Their Graphs*, New Mathematical Library, Vol. 14, Mathematical Association of America, 1964.
17. Klamkin, M., *International Mathematical Olympiads, 1978–1985*, New Mathematical Library, Vol. 31, Mathematical Association of America, 1986.
18. Klamkin, M., *USA Mathematical Olympiads, 1972–1986*, New Mathematical Library, Vol. 33, Mathematical Association of America, 1988.
19. Kürschák, J., *Hungarian Problem Book, volumes I & II*, New Mathematical Library, Vols. 11 & 12, Mathematical Association of America, 1967.
20. Kuczma, M., *144 Problems of the Austrian–Polish Mathematics Competition 1978–1993*, The Academic Distribution Center, 1994.
21. Larson, L. C., *Problem-Solving Through Problems*, Springer-Verlag, 1983.
22. Lausch, H. *The Asian Pacific Mathematics Olympiad 1989–1993*, Australian Mathematics Trust, 1994.
23. Liu, A., *Chinese Mathematics Competitions and Olympiads 1981–1993*, Australian Mathematics Trust, 1998.
24. Liu, A., *Hungarian Problem Book III*, New Mathematical Library, Vol. 42, Mathematical Association of America, 2001.
25. Lozansky, E.; Rousseau, C. *Winning Solutions*, Springer, 1996.
26. Ore, O., *Graphs and Their Use, Random House*, 1963.
27. Shklearsky, D. O; Chentzov, N. N; Yaglom, I. M., *The USSR Olympiad Problem Book*, Freeman, 1962.

28. Slinko, A., *USSR Mathematical Olympiads 1989–1992*, Australian Mathematics Trust, 1997.

29. Soifer, A., *Colorado Mathematical Olympiad: The first ten years*, Center for excellence in mathematics education, 1994.

30. Szekely, G. J., *Contests in Higher Mathematics*, Springer, 1996.

31. Stanley, R. P., *Enumerative Combinatorics*, Cambridge University Press, 1997.

32. Taylor, P. J., *Tournament of Towns 1980–1984*, Australian Mathematics Trust, 1993.

33. Taylor, P. J., *Tournament of Towns 1984–1989*, Australian Mathematics Trust, 1992.

34. Taylor, P. J., *Tournament of Towns 1989–1993*, Australian Mathematics Trust, 1994.

35. Taylor, P. J.; Storozhev, A., *Tournament of Towns 1993–1997*, Australian Mathematics Trust, 1998.

36. Tomescu, I., *Problems in Combinatorics and Graph Theory*, Wiley, 1985.

37. Wilf, H. S., *Generatingfunctionology*, Academic Press, 1994.

38. Wilson, R., *Introduction to Graph Theory*, Academic Press, 1972.

39. Zeitz, P., *The Art and Craft of Problem Solving*, John Wiley & Sons, 1999.